# How to be a Successful Scientist

David Julian McClements
Jake McClements
Isobelle Farrell McClements

# How to be a Successful Scientist

A Guide for Graduate Students,
Postdocs, and Professors

 Springer

David Julian McClements
Department of Food Science
Chenoweth Laboratory
University of Massachusetts
Amherst, MA, USA

Isobelle Farrell McClements
Chemistry and Chemical Biology
Cornell University
Ithaca, NY, USA

Jake McClements
School of Engineering
Newcastle University
Newcastle, UK

ISBN 978-3-031-51401-2        ISBN 978-3-031-51402-9   (eBook)
https://doi.org/10.1007/978-3-031-51402-9

This Springer imprint is published by the registered company Springer Nature Switzerland AG
The registered company address is: Gewerbestrasse 11, 6330 Cham, Switzerland

Paper in this product is recyclable.

# Preface

Working as a research scientist is one of the most satisfying and stimulating careers available. Your efforts may benefit humankind by advancing our understanding of the world, creating new products or processes, or solving complex problems. Going to work is often enjoyable, rather than being a drag. But research scientists need a broad range of skills to be successful, which they normally learn during their graduate studies. Typically, these skills are acquired slowly and in a piecemeal fashion as someone progresses through their graduate program. However, we believe that many of them can be gained much more rapidly and holistically by simply presenting them in a clear and concise manner. The purpose of this book is therefore to provide practical advice on how to be an effective research scientist, including managing your time, designing experiments, publishing scientific manuscripts, giving talks, writing grants, defending a graduate thesis, and much more. We mainly wrote this book with graduate students in mind, but our advice will also be valuable to others involved in scientific research, like postdocs, professors, government researchers, and industrial scientists. Our book has a unique perspective because it is written by three scientists from the same family who are at very different stages of their careers but are all passionate about science and its ability to transform our world. David Julian McClements is a distinguished professor who has been an academic for over three decades. He has written several books, published over a thousand scientific articles, given numerous talks, and mentored many students and postdocs. Jake McClements recently completed his PhD and started a career as an academic. Isobelle Farrell McClements is just at the beginning of her PhD journey. Together, we provide advice from different perspectives that we hope will be useful for those wanting to be more efficient and impactful scientists.

Photograph of David Julian McClements, Jake McClements, and Isobelle Farrell McClements in Northampton, Massachusetts (Summer 2023).

Amherst, MA, USA                                          David Julian McClements
Newcastle, UK                                                           Jake McClements
Ithaca, NY, USA                                          Isobelle Farrell McClements

# Declaration of Generative AI and AI-Assisted Technologies in the Writing Process

During the preparation of this book, the authors used ChatGPT to find relevant information. After using this tool, the authors reviewed and edited the content and take full responsibility for the content of the final publication. We also used AI programs to draw many of the figures in the book (Midjourney and DALL-E).

# Acknowledgments

There are many people we must thank for their value contributions to this book. First, we must thank our families and friends who encouraged and supported us throughout the writing process. We would also like to thank several graduate students who provided valuable comments and suggestions for improving their book, including Xianwei Chen and Minghe Wang.

We would also like to thank the students, postdocs, visiting scientists, staff, and faculty in our departments for all of their support. Finally, we would like to thank everybody at Springer, especially Daniel Falatko, for believing in this project and helping to bring it to fruition.

# Contents

# Chapter 1
# Setting Out on Your Academic Journey

## Starting Your Graduate Career

This chapter is mainly targeted at graduate students beginning their academic journey. Starting your graduate studies is both exciting and scary. It is different from your undergraduate studies, where you mainly focused on taking classes and passing exams (and perhaps partying a lot). Now, you are responsible for your own research project, which requires a whole new set of skills. Some of you may have gained research experience during your undergraduate studies by working in a

professor's laboratory. However, these projects only provide a glimpse into the graduate student experience because of the limited time available. This book is designed to give you an overview of the life of a graduate student and to provide you with some tips and tools that will help you succeed. However, what exactly do we mean by successful? Practically, we mean that you will finish your graduate studies on time with all the skills you need to succeed in the future. This includes creating a *curriculum vitae* that will be attractive to your future employers, whether in academia, industry, government, nonprofit organizations, or other areas depending on your career goals.

You need to do several things soon after you begin your graduate studies. Some of these are practical things, such as finding a place to live, establishing how you will get around, setting up a bank account, registering with a doctor and dentist, and finding out where the local supermarket and other shops are. Others are more directly related to your graduate studies:

- Becoming familiar with your graduate program.
- Getting to know your professor and lab mates.
- Establishing your research topic.
- Developing a good research philosophy.
- Thinking about your career goals.

## Familiarizing Yourself with Graduate Program Expectations

Each department has somewhat different requirements for students to complete their graduate program. It is therefore important to establish precisely what your department requires for you to complete your graduate studies on time. Some of the most common elements in graduate programs include a qualifying exam, course work, teaching experience, scientific research, an oral exam, a written exam, a thesis proposal, and a final thesis defense. Most graduate programs do not contain all of these elements. However, it is important to find out what the expectations are in your department, as well as the timeline for completing each element. This information may be obtained from your department's website, from a student handbook, or by speaking to your professor or graduate program director. If your department does not provide one, we recommend you prepare a checklist containing all the major elements and when they need to be completed. An example checklist is shown in Table 1.1. You can then keep this checklist in your records and check off each element as you complete it.

In addition to being sure to complete the specific requirements of your graduate program, you may also want to set some goals for your research progress. For instance, you may want to write a review article in your first year or two, and then write one or more scientific manuscripts and present several talks in the following years (with the number depending on your research field).

**Table 1.1** Example of a checklist containing the major elements needed to complete a graduate program in the United States. This checklist will vary for different universities and countries

| Task to complete | Notes | Timeline (Time after starting) | Date completed |
|---|---|---|---|
| Course work | Required to pass 18 credits of graduate level classes | 12–18 months | |
| Oral proposal exam | Prepare grant proposal and arrange to have four professors serve on oral examining committee. | 18–24 months | |
| Establish thesis committee | Establish thesis committee. Arrange to have four professors serve on this committee. | 18–24 months | |
| Thesis proposal | Write thesis proposal and give oral presentation to thesis committee. | 1–2 years before finishing | |
| Final defense | Request advisor to send email to graduate program director about results of defense, date passed, and thesis committee member names. | 4–5 years | |

## Choosing a Professor

In some cases, you may be assigned to work with a particular professor, or you may have to work for a particular professor because they are the only person who has funding to support your graduate studies. In other cases, you may have some flexibility in deciding which professor to work with. Therefore, how do you decide? There are several factors to consider, which can be mainly divided into field of study, productivity, environment, and personality (Fig. 1.1):

- *Field of study*: The professors in your department will work in a variety of different research areas. For instance, in a chemistry department, they may work in the general areas of organic, inorganic, physical, or analytical chemistry. They may also use their specific expertise to focus on particular topics, such as renewable energy, sustainable polymers, or drug delivery systems. Ideally, you should choose a topic you are passionate about. The more excited you are about your research, the easier it will be to put in the long hours of work needed over the next few years. You may also want to choose a topic that is compatible with your career goals. If you intend to become an academic, you might want to select a "hot topic" that is on the cutting edge of research and is likely to be important in the future. Universities often want to hire new faculty members who are doing pioneering research rather than those doing more traditional research. Conversely, if you intend to work in industry, you may want to work with a professor who will help you develop the skills needed. For instance, you can get experience with the fabrication and characterization methods typically used in the companies you would like to work for.
- *Productivity*: The research productivity of a professor is another crucial factor to consider when selecting a laboratory to work in. There are enormous variations in research output between different professors depending on their field of expertise, research group size, research philosophy, creativity, and efficiency. Some

**Fig. 1.1** Choosing a professor with a personality, working culture, and productivity that match your needs can facilitate your graduate studies

professors only publish one paper every few years, whereas others publish multiple papers per year. Typically, it is better to work in a highly productive laboratory, as you will be involved in more research projects that lead to scientific publications. This will help to bolster your *curriculum vitae* and make you more competitive when you look for your next job. However, some professors (especially those working at highly competitive universities) focus on publishing high-impact manuscripts, such as those that appear in prestigious journals such as *Nature* or *Science*. These manuscripts often require many years of hard work involving numerous people because the problems being addressed are so complex and at the outer edges of our knowledge. Being part of one of these manuscripts can greatly increase your reputation. However, this is high-risk/high-reward research, and there is no guarantee that it will lead to findings that will be published in one of these prestigious journals. Even so, some people are really passionate about this kind of research and are willing to take the risk. For other people, it may be safer to choose a topic that is more likely to lead to results that will be publishable. In general, it is a good idea to do some background research on the number and quality of scientific papers being published by different professors in your department when selecting the most appropriate one to work with. This can be done using literature database search programs such as Web of Science, Scopus, or Google Scholar. Often, assistant professors who are working toward tenure have the strongest motivation to publish, but they also have to establish their labs and research programs, which can take some time. It may also be possible to obtain some insights into the productivity of a research group by checking the number of papers that the group members have published by the end of their PhD degrees, which can also be done using these database search programs.

- *Culture*: Professors also foster different working environments within their research groups, which can greatly impact your experience during your graduate studies, including your chance to develop the skills you need to succeed. Some researchers in highly successful and well-funded laboratories have enormous lab groups, sometimes with more than 50 people. As a result, it may be difficult to see your professor very often, and so you do not get much direct mentorship. In contrast, other laboratories may only contain a few students and postdocs, so you may have a much better chance of getting to know your professor and receiving more guidance. Professors also vary in the frequency of individual and group meetings that they expect to have with their graduate students. Some professors may hold meetings with students every week or two, whereas others only expect to meet the student when they have a specific problem or question. Some professors actively mentor their students and ensure they develop the knowledge and skills they require for the rest of their careers, whereas others leave it up to the students to work it out for themselves. You should therefore find out what the working environments of the different professors in your department are and then identify one that suits your goals and personality. This can be done by speaking with the professors or the people who work in their laboratories. For instance, you could find the contact information of students working in different professors' laboratories and contact them by email or social media before you arrive in your new department.*Personality*: Professors are people, and like all people, they vary greatly in their personalities. Some professors can be kind, supportive, and encouraging, whereas others can be mean, uncaring, and demanding. Some professors may be relatively flexible and give you time off when you need it, whereas others may expect you to be in the laboratory all the time. Again, you want to find a professor whose personality and work expectations match your needs. You can often get some idea about this by talking to other students and postdocs in the department who have worked with the professor before. Alternatively, you can speak with the professor themselves. You may be willing to work with someone who has a bad reputation if they are a superstar scientist who may advance your career. However, be sure to think about this when deciding who to work with. Poor student-supervisor relationships are one of the main reasons why graduate students fail to complete their degrees, as well as for student mental health problems.

## Getting to Know Your Lab Mates

After starting your graduate studies, you may work in the same department and research group for several years. It is therefore important that you establish a good working relationship with your professors and lab mates (Fig. 1.2). In addition, your lab mates may be people you can hang out with in your spare time. Your lab mates may include technicians, postdoctoral research fellows (postdocs), visiting scholars, PhD students, Master's students, and undergraduate students. For those unfamiliar with the different kinds of people who typically work in academic research laboratories and their different roles, we give a brief overview here (if you already know this, just skip to the next section).

### *Your Professor*

Typically, professors are very busy people who have many duties, as well as supervising and mentoring the members of their research team. They may have to prepare and teach classes, write scientific manuscripts and grants, attend scientific meetings,

**Fig. 1.2** It is important to get to know and develop a good working relationship with your professor and lab mates

serve on departmental or university committees, work with government agencies, consult with industry, and perform various other tasks. Consequently, you should plan your interactions with your professor carefully, so your meetings are efficient and effective. Make sure you are well prepared for each meeting and have specific questions you want to ask. It is a good rule of thumb to have at least attempted to find an answer to your question before asking your professor. Once you have settled into the department, contact your professor (usually by email) and ask them for a first meeting. They may explain the graduate program to you and the expectations required of a graduate student working in their lab. They may also introduce you to the research topic you will work on. Sometimes, this may be a very well-defined topic, but other times they may ask for your input. Therefore, have an idea about what kind of research you might be passionate about working on before your first meeting.

## Your Lab Mates

Technicians are often permanent employees of the university that are responsible for the general management of research laboratories. They help to orientate and train new students, order chemicals, supplies, and equipment, and ensure that the laboratory is clean, safe, and runs smoothly. Postdocs are people who have already completed their PhD studies and are now working for a professor on a specific research topic. They do not need to take any more exams, but they are often highly motivated to perform research that will lead to scientific publications because they may want to become professors themselves. As a result, they want to build a strong resume that will make them competitive when a professor position opens up somewhere. For this reason, working with postdocs is often a good idea because they have a strong incentive to publish scientific manuscripts. Visiting scholars may be students, postdocs, or professors who come from another university, often international, who are working in the laboratory on a specific topic. Interacting with them is a nice way to learn more about people from different countries and cultures, which can greatly enrich your graduate study experience. Graduate and undergraduate students working in your laboratory may be doing projects similar or different from yours. It is crucial to talk to these students to find out what they are doing. Then, you can identify people who can help you with your research and make sure you are doing something sufficiently different from them. If your projects overlap too much, it will be more difficult for you to publish your research independently. PhD students typically work for 3–6 years, Masters students for 1–2 years, and undergraduates for a semester or two. It is advisable to identify the most senior PhD students working in a similar area to you, as they will have the most knowledge and experience and will most likely be able to advise you. For instance, they can help to teach you the different analytical instruments, experimental protocols, and data analysis techniques needed in your own studies.

## Finding a Mentor

Starting a graduate program can be daunting, confusing, and overwhelming. It is therefore a good idea to identify a mentor who has been through the process before because they can provide you with advice and support. This is usually someone in your lab group who is a senior graduate student or postdoc. You could speak to your professor about recommending a suitable mentor for you. You may even want to have a mentor outside the university who can help you with your career development, like a research scientist who works in industry. A mentor can provide practical advice such as where to order chemicals or who to ask to use a piece of equipment in the laboratory. They may also provide you with emotional support if you are feeling overwhelmed or stressed.

## Timing – How Long Will It Take?

The length of your graduate studies can vary considerably depending on your topic, professor, department, location, funding source, and how rapidly your research progresses. Typically, however, a PhD takes approximately 3–6 years. Ideally, you want to complete your studies as soon as possible, then you can move on to the next stage of your career. On the other hand, you want to build a good *curriculum vitae* to help you secure your next position. You may therefore need to publish articles in academic journals, give presentations at conferences, and (when possible) win awards or scholarships. Typically, your research productivity and output increase as you progress through your PhD, so you tend to publish and present more in the later stages of your graduate studies. For this reason, it may be beneficial to stay for a longer time, so you can build a stronger *curriculum vitae*. However, this depends on their being sufficient financial support available to continue your studies. The bottom line is that you want to be as productive as possible throughout your PhD to finish in good time, with as many accomplishments as possible.

## Developing the Skills You Need for Your Future Career

You should use your time as a graduate student to develop the diverse range of skills you will need to be successful throughout your career. It is important to become an expert in your specialized field of study, which means you need to learn the theoretical principles, analytical instrumentation, experimental protocols, and data analysis methods needed in this field. However, numerous other "soft" skills are also important to becoming a successful scientist, which you should also try to develop. These skills include time management, creativity, critical thinking, communication, independence, teamwork, literature searching, and resilience.

## *Time Management*

Throughout your graduate studies you may have multiple tasks to complete, including taking classes, teaching classes, preparing for exams, carrying out research, analyzing data, presenting results, attending conferences, and preparing oral presentations. All of these tasks put demands on your time. It is therefore extremely important to develop good time management skills so you can achieve your goals. First, make a list of all the tasks you need to accomplish. Second, rank the different tasks according to their importance (high/low importance) and urgency (high/low urgency) (Fig. 1.3). Third, look at the time you have available in your calendar. Work on the urgent and important tasks first, leave the unimportant and nonurgent ones to last (or don't do them at all if they are not really necessary). An important skill is learning when to say "no." You may get asked to do a lot of things that are not critical to completing your graduate studies - be prepared to turn them down if necessary. We briefly give some tips for handling different tasks below:

- *Important and urgent*: If possible, do these tasks as quickly as you can. Set aside a block of time, focus, and get them finished promptly. Examples of this kind of task are completing a class assignment or take-home exam, as well as responding to the reviewer's comments on a scientific manuscript.
- *Important but not urgent*: These are important tasks that do not need to be done immediately. Look at your calendar and schedule a time when you can work on these tasks. Examples of this kind of task are writing a scientific manuscript,

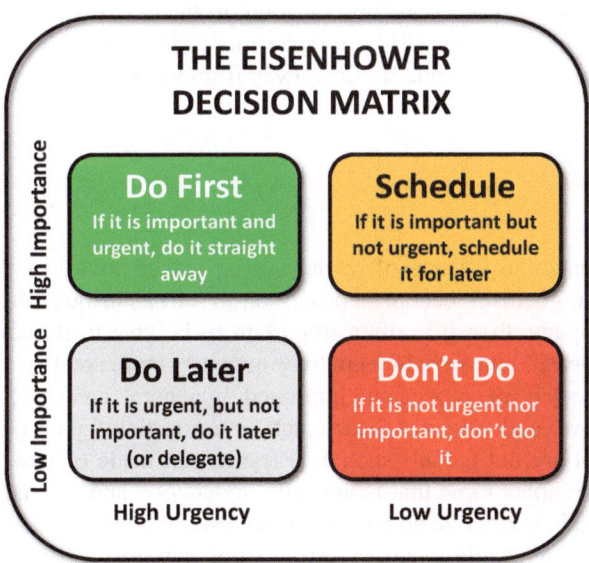

**Fig. 1.3** The Eisenhower decision-making matrix can help with time management and prioritizing tasks

preparing for a presentation at a conference, or writing your thesis. These tasks can be added to a list of things to do and then worked on when you have free time. Then, when they do become urgent, you will already have made some progress on them.

- *Not important but urgent*: Some tasks may not be essential but have a short deadline. First, decide whether you need to do these tasks at all. Turning them down or delegating them to someone else may be possible. If you do need to do them, then work on your important and urgent tasks first, and then work on these ones. As an example, there may be a seminar in your university that is on a subject that is far from your field of expertise. The seminar may be taking place soon, but you have another urgent and important task to complete. In this case, you want to skip the seminar, as it will not significantly impact your graduate studies. However, if you have more time, attending may be beneficial because it would broaden your knowledge.
- *Not important and not urgent*: Some tasks are not important and not urgent. You can often ignore these tasks. This kind of task might be reorganizing the books on your bookshelf to put them into alphabetical order or checking your social media posts every 5 min.

During your graduate studies, you may be overwhelmed with many different demands. Taking time out for yourself is always important to ensure that you maintain your mental health and well-being. You may be able to do some of your academic tasks in a social environment, such as answering your emails or reading scientific papers in a café, which can be more fun than working in the lab all the time. It is also important to take time completely away from work, such as hiking, working out, doing a sport, traveling, or going to the pub, café, or restaurant with your friends. Typically, you will be a much more efficient, effective, and happier researcher if you take some quality time away from work.

## *Creativity*

Important advances in science often depend on the creativity of the individuals involved. If your goal is to become a professor, to start a business, or to work for an innovative company, then it is often important to be able to think creatively and generate new ideas. Some people seem to be naturally more creative than others, but it is a skill that can be developed and improved. In general, creativity involves coming up with novel and valuable ideas (such as a nonstick frying pan) rather than novel ideas that are not useful (such as a frying pan made of ice). The origin of creativity is a complex topic that is not fully understood and depends on both biological and social factors. We all possess some level of creativity because it is hardwired into us through evolutionary pressures – our ancestors were those people who could creatively solve problems in their environment and therefore be more likely to survive and procreate. Creativity also depends on a person's exposure to diverse experiences, cultures, and social interactions throughout life. Some societies and places are more accepting and encouraging of diversity than others, which can stimulate creative thinking. Typically, creativity relies on developing a deep understanding of the area you are working in (as well as other areas) and then entering into a state of mind that allows your unconscious to recombine your knowledge and ideas in novel ways. This means that it is crucial to read many scientific papers and books and to attend scientific talks in your research field and related areas. Often, an idea that is commonplace in another field can be brought into your research field, leading to new advances.

## Critical Thinking

Critical thinking is an important tool to develop during your graduate studies. It involves questioning the assumptions underlying your and others research, carefully assessing evidence to ensure the research was designed, performed, and interpreted appropriately, and considering problems from alternative perspectives. As an example, at the beginning of one's career, it is common to assume that everything you read in scientific articles or hear at scientific conferences is correct. However, researchers, reviewers, and editors can make mistakes, and so it is important to critically assess any materials you are utilizing within your own research. Thinking critically about other people's work helps you to think critically about your own, which can help to prevent you from making silly mistakes.

## Communication

The ability to communicate with different audiences in oral, visual, and written formats is critical to being a good scientist and advancing your career. There is no point doing important scientific work if nobody finds out about it. You should therefore be sure to develop good communication skills throughout your graduate studies. The best way to do this is by writing scientific papers, preparing posters, and giving talks at lab meetings, classes, and conferences. There may be resources at your campus that can help you develop effective communication skills, such as career development centers. There are numerous books on developing good speaking and writing skills (see resources at the end of this chapter).

Writing is a critical skill for scientific researchers because most science is communicated through manuscripts and published in scientific journals. You should therefore pay particular attention to developing effective writing skills. Your writing should be informative, engaging, and concise. Tips on writing scientific manuscripts and developing good writing skills are given later in this book (Chap. 6). Other forms of communication are also important for academics, including speaking and graphics. You may be asked to present your research findings in the form of a talk or a poster at lab meetings, scientific conferences, industrial events, or as part of your exams. Tips for preparing oral presentations and posters are also provided later in this book (Chap. 9). Developing an engaging speaking style can help get you and your research recognized by other scientists in your field, who may be potential future employers. If they are impressed with the quality of your talk, they are more likely to hire you in the future.

Having good graphics skills is also essential for effective and impactful scientists. Many scientific journals require authors to submit a graphical abstract with their manuscripts, a single image that captures the essence of the research. A high-quality image can attract readers to your work and increase its impact (Fig. 1.4). Moreover, some scientific journals invite authors to submit a visually engaging image that may be used as the cover of the journal, further increasing the impact of your work. Good images can also greatly increase the comprehension and impact of your scientific papers, posters, and talks, enhancing your visibility in your field of expertise. Again, this helps increase your scientific reputation and may help you get your next job. There are different kinds of computer programs that can help you

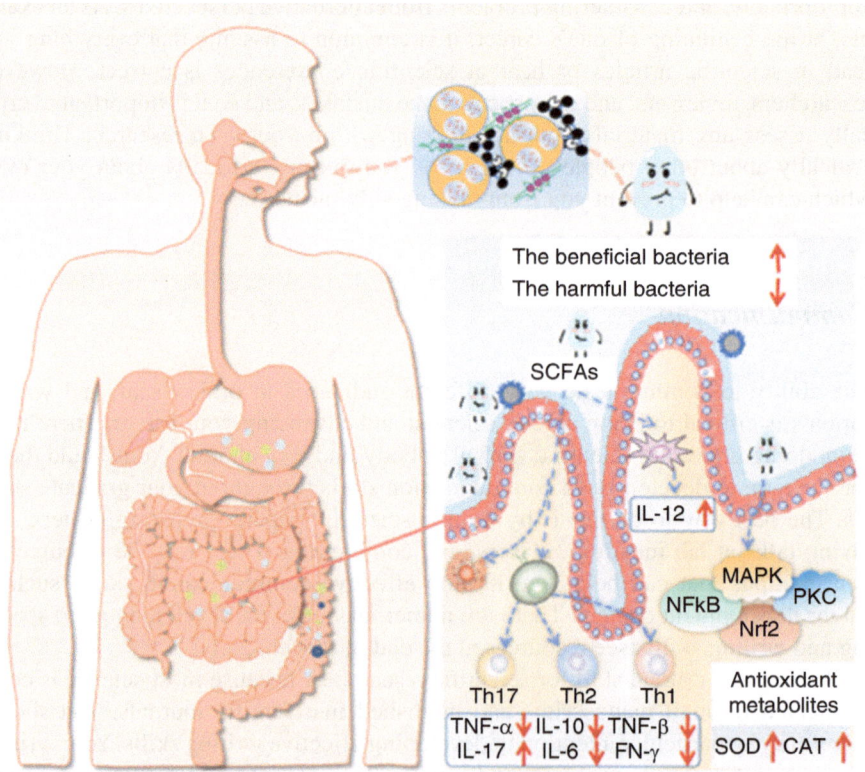

**Fig. 1.4** Having the ability to create engaging and informative images can greatly enhance the impact of your scientific research. (Image from Gu and coworkers (with permission) (Gu et al. 2022))

create visually impactful images, such as Microsoft PowerPoint, Adobe Illustrator, or Adobe Photoshop. In addition, artificial intelligence programs (such as Midjourney) are becoming increasingly good at creating images, such as some of those shown in this chapter. It is advisable to become familiar with one or more of these graphical design programs to increase the impact of your research. In the future, other forms of communication may become more important, such as online videos or podcasts.

## Independence

If you are pursuing a graduate program, you will likely become a future leader in your field. For example, you may become a professor, a manager, an entrepreneur, or a government official. It is therefore critical that you develop the skills needed to think and act independently. By the end of your graduate studies, you should be able to come up with original ideas, design and perform experiments, analyze the results, and communicate your findings. This may seem daunting when you start your graduate studies, but you should actively work to develop independence as you progress through your degree. This may be achieved by brainstorming new ideas with your professor or lab mates, helping to design experiments yourself, writing the first draft of scientific manuscripts, helping to revise manuscripts after they have been reviewed by academic journals, and volunteering to give scientific presentations.

## Teamwork

Scientific knowledge has expanded enormously over the past century, and it is often difficult for a single individual to make transformative advances on their own. Many areas of science are extremely complicated and benefit from multidisciplinary teams that involve numerous people with complementary skills to solve complex problems. It is therefore important for you to be able to identify situations where a team approach would be beneficial to solve a problem, as well as to identify potential collaborators, put together a team, and work with other people effectively. This may often be the main responsibility of a professor leading the research, but graduate students may also be involved as part of the team. Several factors are important to successfully working in teams. You need to ascertain precisely what your role is in the team and what is expected from you. You should make sure you finish any tasks assigned to you on time, so you are not holding up other people in your team. You also need to respect everybody else's time – don't hold meetings or ask for detailed information unless it is absolutely necessary. You need to have an open mind when listening to other people's perspectives but also be prepared to promote your own perspective. People working in different scientific disciplines have different knowledge, approaches, and skill sets. It is crucial to appreciate this and be willing to learn from others. One of the most important things is to be a nice person. Don't be rude, domineering, or dismissive. Effectively working in teams is becoming increasingly important in science, so developing these skills during your graduate studies is important whenever possible.

## Literature Searching

Another critical skill for scientists is to be effective at searching the literature. In the past, this used to involve spending hours in the library poring over thick volumes of books containing lists of keywords from scientific articles that had to be cross-referenced and physically searched for on the library's shelves. Now, this can be done much more quickly and comprehensively using internet search programs such as Google Scholar, Scopus, or Web of Science. In addition, the widespread availability of artificial intelligence-based search routines is growing rapidly, providing powerful tools to rapidly discover knowledge about a particular subject. Nevertheless, it is still important to come up with good keywords and questions to find the information you are looking for.

For several reasons, it is critical to carry out a thorough literature search at the beginning of your graduate studies. First, it helps to establish what has already been done in the area you intend to work in. You don't want to do years of research and then discover somebody has already done it and published it. Ideally, you want to identify gaps in existing knowledge where you can focus your research efforts and advance the field. Second, reviewing the literature helps to identify the analytical instruments, experimental protocols, and data analysis methods typically used in the area you intend to work in. Third, the knowledge you gather may be helpful in writing the introduction section of your thesis and scientific manuscripts. Fourth, this knowledge may also be used to write a review article in your area of research, which can help to establish your reputation in the field.

## Laboratory Practice

You may work in a laboratory with numerous other lab mates. It is important that you work safely, efficiently, and effectively. Typically, you will need to carry out a health and safety course before you are allowed to work in the laboratory, which is usually organized by your university. You may have to repeat this course annually to be sure you are up to date with best lab practices. Working efficiently means that you should not waste time or resources - chemicals and reagents are often expensive to purchase and dispose of – and so you should minimize the number of experiments you perform that lead to little or no results. The most likely cause of wasted time and resources is bad planning. If you do not plan carefully, you may spend days, weeks, months, and sometimes even years doing experiments that are of no

**Table 1.2**  A checklist of the critical skills needed to be a successful scientist. You might want to rate yourself for each skill throughout your graduate studies and work on the skills you think you need to improve

| Skill | Description | Rating[a] |
|---|---|---|
| Creativity | Ability to come up with new ideas | |
| Critical thinking | Ability to critically evaluate your and others' research | |
| Time management | Ability to effectively manage all the tasks you need to carry out | |
| Oral communication | Ability to effectively communicate your research findings orally | |
| Written communication | Ability to effectively communicate your research through written scientific publications | |
| Graphics skills | Ability to draw informative and visually impactful images | |
| Independence | Ability to work on your own | |
| Teamwork | Ability to work with others | |
| Lab practice | Ability to work efficiently and safely, while keeping the laboratory clean and tidy | |

[a]Rating scheme: Strong, Fair, Needs improvement

use. This means you have wasted your time, wasted your professor's funds, and will take longer to graduate. Always carefully design all your experiments before going into the lab to perform them. A few hours planning can save you months of wasted time. It is also important to keep your laboratory space and all of the equipment you use clean and tidy. If somebody else is starting an experiment, they don't want to have to clean up after you first. Moreover, keeping everything clean and tidy reduces any potential safety issues, as well as the possibility that the equipment may wear out and breakdown. Critical equipment that breaks can take a lot of time and money to repair, which can hold up your research, as well as everybody else's working in your lab (making you very unpopular). You should therefore always report any equipment that appears to be faulty to the lab technician or professor so they can get it fixed promptly.

We would advise you to make a checklist of the most important skills you need to develop during your graduate studies and then rate yourself (Table 1.2). If you need help or advice on developing a particular skill, you may be able to work with your professor or lab mates, or there may be specific resources available at your university or online that can help you develop them.

## Thinking About Your Career Goals

Even if you are just beginning your graduate studies, it is important to be thinking about your ultimate career goals. Do you want to work in academia, industry, government, or a nonprofit organization? Do you want to be an entrepreneur who establishes their own company? Do you want to work as a research scientist, manager, educator, or sales representative? By thinking about your career goals at the beginning of your graduate studies, you can tailor your program to develop the skills you need and build a *curriculum vitae* that will help you secure your next job. For instance, if you plan to become an academic, then it may be important to actively seek out opportunities to write scientific papers, present at scientific conferences, give lectures, and write grants. Alternatively, if you want to become a manager or an entrepreneur, you may want to take some management or business classes as part of your degree. If you want to work in research and development in industry, you may want to become familiar with a broad range of analytical methods commonly used in that industry. Talk to your professor about your career goals so they can give you advice on the best way to achieve them.

## Developing the Right Attitude

Developing the right attitude toward your research will help you throughout your graduate studies. People with a positive attitude, who think critically about their research and are committed to finding solutions to their problems, are typically more successful than those who are negative, mindlessly carry out their research, and give up when they encounter any problems. We all have different personalities that have developed throughout our lives. However, whenever possible, it is important to develop a positive problem-solving mindset when embarking on graduate research. There will always be challenges to overcome in your research – if it was easy to do, your professor would have a technician do it. These challenges may be complex scientific problems that are difficult to solve or simple practical problems, such as equipment breaking, chemicals not being available, or your experiments being spoiled by someone else's carelessness. To solve complex scientific problems, you often have to be extremely creative and tenacious. Your professor may be unable to give you a simple answer, so you may have to do a lot of research and background reading to solve the problem. Of course, knowing when to give up trying to solve a problem is also important. Perhaps it is simply not solvable with the equipment and resources available to you, and it would be better for you to move on to something else. Again, having a frank discussion with your professor can help you decide when it is important to keep pursuing a problem and when it is better to move on.

## Work-Life Balance and Mental Health

Working as a graduate student is often grueling. It may involve long hours in the laboratory, often coming in on nights or weekends to complete experiments, with little vacation. Moreover, the research may go differently than planned. The equipment you need may be broken, the chemicals you need may be unavailable, your experiments may not go as anticipated, or your computer may get a virus and you lose all your data. This can be extremely frustrating, discouraging, and sometimes depressing. It is therefore important to consider your mental health and to establish a good work-life balance. Try to establish a good friend group that can support you.

Go out together for coffee, drinks, hikes, or whatever else you find provides a break from thinking about or doing research all the time. If you are feeling depressed and it is affecting your work and life, there are usually resources on campus that can help you. You should also speak to your professor who can support and advise you.

## Adapting to Your New Environment

Unless they have completed another degree at the same university, most graduate students are new to the place where they are working. They may come from a diverse range of geographical locations, which may be in the same country or a different one. For international students, the language commonly spoken in their new location may be different from their native tongue, which can lead to problems in understanding and communicating, especially at the beginning of the graduate program. Universities typically have resources to help students become more fluent in written and spoken language. In addition, it is a good idea for new graduate students to refine their language skills by watching or listening to programs on the TV, radio, or internet in the local language. If you come from a country that has many other students working in your new university, there may be a student group with people in a similar situation. The customs in your new location may differ from those you are familiar with, and it may take some time for you to adjust to them.

Starting your graduate studies may also require you to adopt a new approach to your work. As an undergraduate, someone may carefully guide your research and be there to help you when things go wrong, but in graduate school, you have to learn to be much more independent and to solve problems yourself. This can take some getting used to. You are likely to make some mistakes as you adopt to this new way of working, but this is completely normal and should not discourage you. Failure often leads to a better understanding of a problem and to the development of new insights and skills.

## Concluding Remarks

Starting a graduate program can be overwhelming. You have many new skills to learn, you are working in a new environment with unfamiliar people, and you are expected to produce an original body of research by the end of your graduate studies. However, it is also a very exciting time when you can grow as a person and a scientist, meet new people (often from all over the world), learn interesting new things, and create new knowledge that may make the world a better place. In the remainder of this book, we provide advice about the most important challenges you will encounter during your graduate studies, such as planning your research, writing scientific manuscripts, presenting scientific talks, and defending your thesis. We hope these tips will be useful, and you will enjoy the often grueling but rewarding life of a graduate student.

**Key Points**
- The focus of your graduate studies is to become an effective independent researcher.
- Each graduate program has different expectations – make sure you know yours.
- Choose a professor who is productive and has a personality compatible with yours.
- Get to know your lab mates and develop a good working relationship.
- Find a good mentor who can provide academic and emotional support.
- Organize your time to complete your studies on time while building a strong CV.
- Develop strong time-management, creativity, critical thinking, communication, and networking skills.
- Tailor your graduate studies to develop the skills needed for your career goals.

## Resources

Anderson S (2016) Ted talks: the official TED guide to public speaking. Mariner Books, Boston

Gastel B, Day RA (2022) How to write and publish a scientific paper, 9th edn. Greenwood, Santa Barbara

Gu QZ, Yin Y, Yan XJ, Liu XB, Liu FG, McClements DJ (2022) Encapsulation of multiple probiotics, synbiotics, or nutrabiotics for improved health effects: a review. Adv Colloid Interface Sci 309

Olson R (2018) Do not be such a scientist: talking substance in an age of style, 2nd edn. Island Press, Washington, DC

# Chapter 2
# The Scientific Method: A Knowledge Machine

## What Is Science?

It may come as a surprise to many practicing scientists, but there are still heated debates amongst historians, sociologists, and philosophers about what science is and how it works. Scientists tend to just do science. What makes it so difficult to define "science", is that scientists do so many different activities as part of their work, using a broad range of approaches. Some scientists try to understand the origin of the Universe by looking into space using powerful telescopes. Others try to elucidate the fundamental structure of matter by smashing atoms together and looking at the fragments formed. Some aim to understand the biochemical processes that underpin life by probing the complex molecular events that occur inside cells. Others try to understand how our brains work by monitoring the electrical activity

D. J. McClements et al., *How to be a Successful Scientist*,
https://doi.org/10.1007/978-3-031-51402-9_2

of neurons. While still others work to enhance the efficacy of anticancer drugs by using nanotechnology to deliver them to specific parts of the body. These are just a few examples of the vast range of problems scientists work on. A truly comprehensive list would be enormous, which is not surprising given that scientists are trying to understand, predict, and control the vast, complex, and dynamic world we live in. Scientists working on different problems use different methods to collect, analyze, interpret, and present data. Despite this, several features of science distinguish it from other areas of human activity (although there are many overlaps):

## Subject Matter

Science aims to provide fundamental insights into the nature of our world. What is the world made of? How are its constituents structured? How do things interact with each other? How do things change over time? The natural sciences (like physics, chemistry, and biology) mainly focus on the physical world, whereas the social sciences (like sociology, psychology, and anthropology) mainly focus on human behavior and interactions. However, there is often overlap between different scientific disciplines, such as trying to relate neuronal or hormonal activity to human behavior. Indeed, some of the most important scientific advances are made at the intersection between different disciplines.

## Abstraction

Scientists cannot understand how the entire world works due to its vastness and complexity. For this reason, most scientific research focuses on a specific well-defined aspect of the world that is simpler to study and understand. This may be the behavior of a beam of neutrons in a particle accelerator, the functioning of a specific protein in a mammalian cell, the mating rituals of a particular species of spider, climate change in the Antarctic over the past 100 years, or the evolution of a distant galaxy. An effective scientist knows which aspect of the world to focus on. It should be simple enough to understand, but not so simple that it does not provide any useful advances in knowledge. Moreover, it should be a significant problem. Ideally, it may lead to an advance in our fundamental understanding of the nature of the world, or it may be of practical importance to humanity.

## Objectivity

Another critical aspect of science is that it aims to be as objective as possible. We all have our own preferences and biases that may influence the way we design, perform, and interpret our studies. For instance, if someone believes they have found a new drug that may cure people with Alzheimer's disease, they may be more prone to discard data that does not agree with their expectations than data that does, leading to unreliable and biased results. It is therefore always important to be aware of any potential conscious or unconscious biases that might affect your studies, and where possible, eliminate them. Science has several mechanisms to avoid bias, such as the need to have your work peer-reviewed by other scientists before it is published, and the fact that other scientists can read your study and try to replicate it. Science has developed a culture of objectivity over the last few centuries. In scientific publications and presentations, there is no room for interpretations based on authority, religion, politics, philosophy, or personal preferences. This approach has proven to be an extremely effective means of advancing our fundamental understanding of the world. The scientific revolution that occurred during the sixteenth and seventeenth centuries marked a major change in the way humans gained knowledge about how the world works and it has led to most of the advances in our standard of living since then.

## *Data-Driven*

Science is usually based on collecting empirical data from carefully designed studies. This data may come from a variety of sources, including observations, surveys, experiments, and simulations. For example, it may come from measurements of the temperature in a particular region over time, questionaries of peoples' eating habits and health status, measurements of the effects of a crosslinking agent on the mechanical strength of plastics, or computer simulations of the climate in Sub-Saharan Africa. It is important to decide what is the most appropriate data to collect and how to analyze and report it. In science, it is also important to specify how the data was obtained and analyzed so that other people can replicate the study and assess its value.

## *Hypothesis-Driven*

Scientists typically develop hypotheses that are tentative explanations or predictions about some aspect of the world. These hypotheses should be clear, concise, and testable. The validity of a hypothesis is then established by carrying out a carefully designed study, which may involve observations, experiments, or surveys. The predictions of the hypothesis are compared to the empirical data. If they agree, the hypothesis is accepted, but if they do not agree, it is rejected. If rejected, a hypothesis can be reformulated based on the new knowledge obtained and then tested in another study. The development, testing, and refinement of hypotheses is one of the most important ways that scientists advance our understanding of the world.

## *Mathematical Modeling*

Many areas of science use mathematical models to understand, describe, or predict some aspect of the world. These models usually include one or more mathematical equations. In some cases, these equations are derived from fundamental physical, chemical, or biological principles, like the laws of motion, the theory of gravity, quantum mechanics, or natural selection. A well-known example of this type of equation is Einstein's $E = mc^2$, which relates energy to mass, but there are many more equations commonly used in other scientific fields that have been derived from first principles. In other cases, the equations used to describe the world are not derived from fundamental scientific principles. Instead, they are used because they have a mathematical form that is suitable for describing some event in nature. For instance, the eq. $C(t) = C_0 \exp(-kt)$, describes the observation that the concentration ($C$) of a substance undergoing a chemical reaction often decreases exponentially over time ($t$) from its original value ($C_0$) at a rate that depends on a constant ($k$). This type of equation may not provide any fundamental insights into the origin of the processes occurring, but it does provide a convenient means of comparing the rates of the reaction under different conditions (Fig. 2.1). Mathematical equations may also be used as part of the *Artificial Intelligence* (AI) algorithms used to establish correlations between different phenomena and make predictions about the future. Again, these models may not provide any fundamental insights into the origin of the observed events, but they still generate useful information about the world. The suitability of a mathematical model for understanding, describing, or predicting some aspect of the world can be tested by carrying out studies and comparing the predictions of the model to the empirical data collected. If the model does not work, then

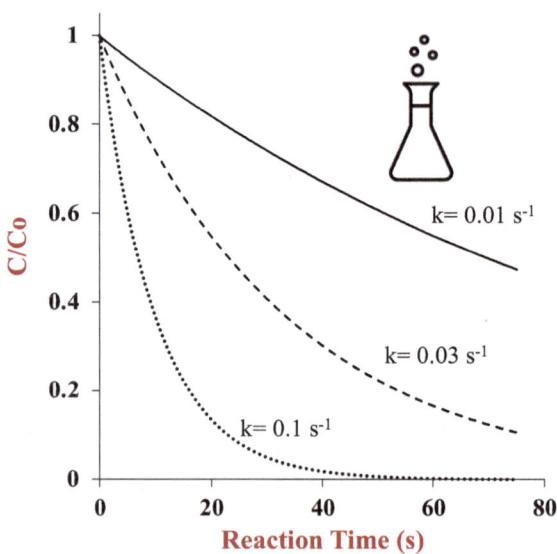

**Fig. 2.1** Some types of mathematical models are simply used to describe (rather than understand) certain aspects of the world. For instance, the eq. $C(t) = C_0 \exp(-kt)$ can be used to describe how the concentration of a substance decreases over time during a chemical reaction, which can be characterized by a rate constant (k)

it may need to be refined or discarded. Another important feature of science is therefore the derivation, testing, and revision of mathematical models.

## *Logical*

Science is typically characterized by an emphasis on logical consistency. A scientific hypothesis or theory should be logical and not contradict the laws of physics. Conclusions should follow from premises based on the laws of logic such as the laws of identity, contradiction, exclusion, and sufficient reason.

## *Rigorous*

Science is characterized by an emphasis on rigor. Hypotheses should be carefully formulated. Studies should be carefully designed, performed, and interpreted. Oral and written presentations should be clear and concise. If your approach to your work is slapdash, then your findings will be unreliable.

## *Reproducible*

In many areas of science, it is possible to design highly reproducible experiments. If all the experimental conditions are kept constant, and the methods used to collect and analyze the data are the same, then similar results should be obtained regardless of who is carrying out the experiment and where they are located. Another important feature of science is that when you report your experimental findings, you should always present them in a manner so that someone else can reliably repeat your work. Whenever possible, scientific experiments should therefore be carefully designed so that they are reproducible. However, in some areas of science, this is simply not possible. For instance, climate scientists can only observe an event happening once in the environment – they cannot turn the clock back and run the same climate event again.

## *Open-Minded*

A good scientist should always be open-minded. Facing queries, criticism, and rejection is commonplace in science and although it can be painful, it is vital for ensuring scientific rigor and quality within the field. Therefore, scientists should be open to criticism and willing to change their views if the evidence supports it.

Having said this, when you do have a strong belief in your ideas, you should present them clearly and rigorously (but politely) argue for them. You can then obtain constructive criticism of your ideas, that will allow you to either strengthen, refine, or discard them, thereby leading to important advances in our understanding of the world.

In this chapter, we begin by considering the nature of the scientific method that scientists use to obtain knowledge about the world. We then consider some of the different sources of data scientists use in their research, including observations, experiments, theories, and simulations, as well as some of the skills successful scientists need at different stages of their studies, such as creativity, rigor, objectivity, and flexibility. Some of the most important contributions of the intellectual pioneers who helped to establish and refine the scientific method are then discussed, including Francis Bacon, David Hume, Immanuel Kant, Karl Popper, and Thomas Kuhn. Knowledge of the contributions of these influential philosophers, sociologists, and historians is important in itself, but it can also help practicing scientists hone their scientific skills. Finally, the utility of transformational/incremental, reductionist/holistic, and trial-and-error/black-box/fundamental approaches of gaining knowledge about our world are highlighted.

## The Scientific Method

One of the most important reasons that science has been so successful at improving our understanding of the world is that it uses a systematic, rigorous, and objective "knowledge-generating machine" known as the scientific method, which can be divided into several stages (Fig. 2.2):

- *Research question*: Initially, scientists focus on a particular aspect of the world and then come up with a specific research question to address.
- *Literature search*: They then do a comprehensive literature review to determine what is currently known about the research question.
- *Hypothesis generaton*: Based on this knowledge, they formulate a clear and concise hypothesis that can be empirically tested to determine whether it is valid or invalid.
- *Observations or experiments*: Scientists then make observations or perform experiments to generate data that is collected, analyzed, and presented.
- *Hypothesis testing*: The predictions made by the hypothesis are compared to the empirical data. If there is good agreement, the hypothesis is valid, but if there is not, it is invalid. The results of the study can then be published in a scientific manuscript, presented at a scientific meeting, and/or included in a thesis. If the original hypothesis is well formulated, it is important to communicate the results even if it is shown to be invalid, as this provides new knowledge about the world.

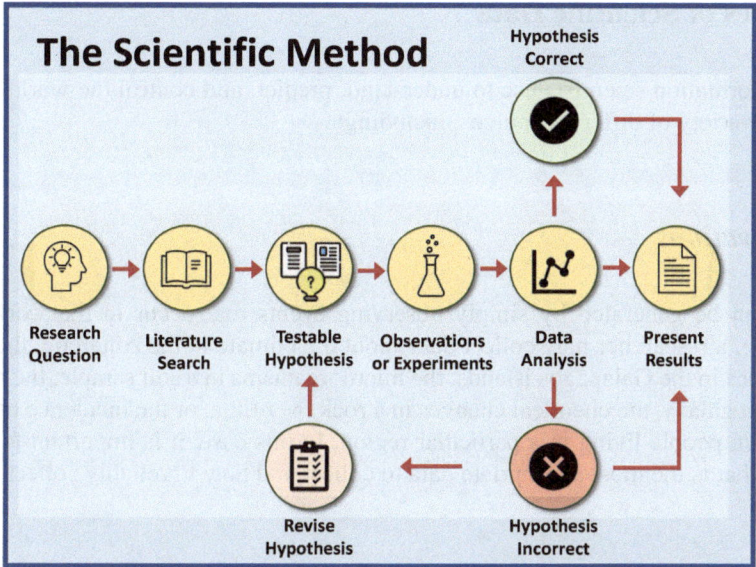

**Fig. 2.2** The scientific method consists of developing good research questions, performing a thorough literature search, formulating a testable hypothesis, carrying out observations or experiments, analyzing the data, testing the hypothesis, and presenting the results. If the hypothesis is found to be incorrect, it can be revised and retested

- *Hypothesis reformulation*: If the original hypothesis is invalid, it can be reformulated based on any new information generated during the study. This new hypothesis can then be tested using the same approach.
- *Theory development*: Once a hypothesis has been systematically tested and shown to be robust under a wide range of circumstances, it might become a theory. A scientific theory is an explanation of some aspect of the world that has been rigorously tested against empirical evidence, such as experiments or observations, and shown to be valid over a specified range of conditions. Typically, a theory can explain previous phenomena, as well as make predictions about the future, which can then be tested through further studies. The theory may be modified or revised if its predictions do not agree with new empirical evidence, leading to new theories with greater explanatory and predictive power.

The scientific method is therefore an iterative process that enables us to progressively improve our understanding of the natural world by generating data and postulating, testing, and refining our hypotheses and theories. The rigorous application of this method is one of the main reasons for the rapid advance in our understanding of the world over the past few centuries. Having said all this, the simple representation of the scientific method described here is highly idealized, and the methods actually used by practicing scientists are often much messier, as will be highlighted later.

### Sources of Scientific Data

The information scientists use to understand, predict, and control the world comes from a variety of different sources, including:

### *Observations*

Data can be generated by simply observing events that occur in the world. For instance, a researcher may collect data about the climate in the Antarctic, the types of finches in the Galapagos islands, the microorganisms in a soil sample, the stars in a distant galaxy, the chemical changes in a rock over time, or the incidence of colon cancer in people living in a particular region. In this case, it is important to determine what is the most appropriate data to collect and how to reliably collect it.

### *Surveys*

Some scientific disciplines, usually those involving understanding human behavior, rely heavily on data obtained from surveys or questionnaires. For example, people may be asked a specific question about their eating habits and health status, and then this information is used to determine if there is a correlation between their diet and health. In this case, it is important to consider what kinds of questions to ask and how to formulate them, as well as who will be your participants. It is also critical to account for any potential biases you may have when developing and administering surveys and questionnaires.

### *Experiments*

Carefully controlled experiments are critical to many scientific disciplines, especially those in the natural sciences, like physics, chemistry, and biology. Typically, these experiments are based on establishing a model system that represents some feature of the world you are interested in, but that is simpler to control, characterize, and understand. For instance, when trying to understand the factors impacting the resilience of an agricultural crop to climate change, you may grow the seeds under standardized conditions (such as soil type, temperature, moisture levels, humidity, and light exposure) and measure the health, yield, and nutritional quality of the crops produced. This will provide more experimental rigor, control, and reproducibility compared to observing the growth of the plant in its natural environment, where the soil and weather conditions may vary greatly. Therefore, you can design experiments that quantify the influence of specific factors on crop growth (such as light, temperature, or moisture levels). It is important to carefully design experiments to produce reliable and reproducible data that other people can replicate.

## Theories

Important scientific insights can be obtained by developing theories that model certain features of the world and then finding out how close the predictions made by these theories agree with reality. As mentioned earlier, theories often consist of one or more mathematical equations that have been developed based on fundamental physical, chemical, or biological concepts. For instance, Newton's laws of motion and gravity can be used to develop equations that predict how planets move around the sun or how fast an object falls to the ground.

## Simulations

Valuable insights into how our world works can be obtained by using simulations performed on computers. Computer simulations are based on creating virtual worlds that model some aspect of the real world, usually in a highly simplified form. They typically require mathematical equations that describe the physics, chemistry, and/or biology of a particular material or phenomenon, as well as some input data about the system being studied. For example, computer simulations of the weather rely on mathematical equations that describe atmospheric processes like changes in air movement (winds), water properties (evaporation, precipitation, melting, and crystallization), temperature (warming and cooling), and pressure changes (rising and falling), as well as empirical information about current windspeeds, temperatures, and pressures in a specified geographical region. The simulations can then be used to predict weather patterns in the future. Computer simulations of materials (like metals, plastics, or concrete) use mathematical models that describe the interactions and movement of the atoms, molecules, or particles they contain, as well as information about the nature of these entities (such as their size, shape, charge, and chemistry). The simulations can help to identify what are the most important aspects of a system that impact its behavior and properties.

## Search Engines

An enormous amount of information about our world has already been generated and published. Moreover, our knowledge is rapidly expanding due to the concerted efforts of scientists and others around the world. Important discoveries can be made by simply searching the existing literature and finding out what is already known, and how different findings are related to each other. There are increasingly powerful search engines, often driven by AI that can be used to find this information and use it in your own research. Whenever you start a new project, it is always important to carry out a thorough literature search to establish what is already known about the subject. This helps you find key information, as well as identify where there are still scientific controversies and gaps in the knowledge. You can then use these insights to guide your own research – especially, to identify what are the most important questions to work on. Moreover, these search engines can also be used to find published information that can be used to develop or test your hypotheses.

One of the important characteristics of a good scientist is being able to ask good questions and identify the most appropriate sources of information to answer these questions. The design, performance, analysis, and reporting of scientific data are comprehensively discussed in later chapters of this book that deal with more practical aspects of the scientific process.

## Key Scientific Skills

Successful scientists employ a broad range of skills during their studies. Here, we highlight some of the most important skills, including those that are most important during the creative, research, and communication phases of a study.

### *The Creative Phase*

The initial phase of a research project often requires a scientist to be highly creative. The scientist must come up with an original question or problem to work on and identify the most effective way of tackling it (Fig. 2.3). Before they can do this, they must have developed a deep and broad knowledge of the intended area of study. Reading books, papers, and other sources of information is vital to become familiar with the concepts, methods, and controversies in the area before starting the idea generation phase. Knowledge of non-related scientific disciplines can also be beneficial at this stage because it can help to introduce concepts and methods that are not normally used in the field of study. Once a scientist has established a good background knowledge, they can begin the process of generating ideas that may be relevant to the problem they are trying to address. This may be done alone or in a team. Working in a team can improve the creative process because each individual brings their own background experiences and perspectives. In this stage, it is important to be open-minded, flexible, and innovative. After generating a range of ideas, the advantages and disadvantages of each idea are critically assessed and the vast majority of them are discarded (Fig. 2.3). At the end of this idea-filtering

**Fig. 2.3** In the creative stage, a scientist or team of scientists generates numerous ideas when they are trying to identify a suitable problem to work on and find potential solutions. The merits of each idea are then critically assessed and the most useful one is selected and formulated into a hypothesis that can be tested

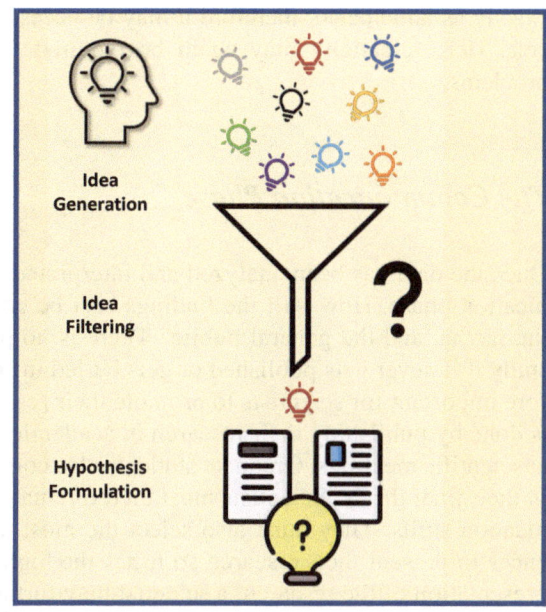

process, a scientist may come up with one idea that they believe is going to be the most suitable problem to work on. This idea should generally be innovative, which will make it easier to significantly advance knowledge and thus publish results arising from the research. The selected idea can then be formulated into a hypothesis that can be tested. For instance, if you came up with the idea that spiderwebs could be used to create biodegradable and sustainable packaging materials, then your hypothesis might be "We hypothesize that spiderwebs can be assembled into thin films with optical, mechanical, and barrier properties similar to those of plastic films."

## *The Research Phase*

Once a scientist has identified an original idea that is innovative and important, and they have formulated a clear, concise, and testable hypothesis, then they can move into the research phase. In this phase, the scientist adopts a much more objective, rigorous, and meticulous approach. Experiments are carefully designed and performed, and the resulting data are carefully recorded, stored, and analyzed. The aim is to be as systematic and objective as possible – rigorous scientific protocols should be established and followed for all the research activities involved in the study. Nevertheless, problems often arise when performing research – things rarely go

exactly as anticipated, therefore it may be necessary to revise the scientific protocols. Here, creativity may again be required to identify and effectively solve
problems.

## *The Communication Phase*

Once the data has been analyzed and interpreted, the scientist enters the communication phase. How will the findings best be communicated to other scientists,
the media, and the general public? There is no point in carrying out a scientific
study if it never gets published or gets buried in some obscure journal. It is therefore important for scientists to promote their research findings. Traditionally, this
is done by publishing their research in academic journals or by speaking about it
at scientific meetings. Graduate students also communicate their research findings
in their final thesis. Scientists must therefore have good writing and oral communication skills. They must also select the most appropriate journals and conferences to present their research so it has the highest impact. After publication or
presentation of the research, a scientist may further promote the dissemination of
their findings to a wider audience using suitable social media outlets (like
LinkedIn, X, or YouTube). In this phase, a scientist should be an avid promoter of
their research, with the aim of increasing its impact. For instance, the researcher
may send any interesting new findings to the press office at their university so they
can write a press release that might be picked up by the popular media. As a result,
it may appear in newspapers, magazines, radio shows, television shows, or
podcasts.

## Pioneers of the Scientific Method

Many practicing scientists never take a course in the philosophy of science. They
are therefore unfamiliar with ideas about the nature of scientific knowledge, how it
is acquired, and what are its limitations. Philosophy of science is typically written
by philosophers, rather than practicing scientists. As a result, it is often written in a
language and style that is difficult for many scientists to comprehend and uses concepts and terms they are unfamiliar with. Moreover, it typically focuses on "big
picture" questions like what can we know, how do we know it, and how can we be
sure what we know is true, rather than on practical questions like what kills this species of bacteria, why is the sky blue, or why is cheese harder than yogurt. Even so,
an understanding of the philosophy of science can enhance the research skills of
practicing scientists.

## *The Nature of Science*

Philosophers have debated for hundreds of years about what science is and how to define it. Practically, however, definitions of science can be given that are acceptable to most scientists. For instance, the Science Council (sciencecouncil.org) gives the following definition:

> *Science is the pursuit and application of knowledge and understanding of the natural and social world following a systematic methodology based on evidence.*

In his book, *The Knowledge Machine* (2020), Michael Strevens highlights that the modern scientific method arose in the seventeenth century when people decided that *only* empirical evidence counts in providing insights into how the world works. Before then, philosophers did value empirical evidence, but they also valued other sources of knowledge and understanding, such as authority, religion, or aesthetics. By being forced into only considering empirical evidence, scientists had to focus on carrying out detailed experiments and observations, which provided a much richer and more comprehensive understanding of our world. A hypothesis or theory was only accepted if it was consistent with objective empirical evidence, not because it was consistent with a holy book, political system, or aesthetic principle.

In the following sections, we present a concise summary of notable individuals who have significantly shaped the modern scientific method. Although many of these individuals did not make major scientific discoveries themselves, they did create the framework that made these kinds of discoveries possible. Additionally, we emphasize the practical relevance of comprehending their contributions for researchers actively engaged in scientific inquiry. It should be noted that these individuals are all white males, which is mainly due to socio-cultural factors throughout history that have limited the access of women and minorities to education, intellectual pursuits, and positions of influence. We must all do our best to recognize, challenge, and address these issues so we can create a society that is more inclusive and equitable. Having a more diverse range of perspectives will also foster important advances in science.

## *Francis Bacon*

Francis Bacon (1561–1626) was an English philosopher who made major contributions to the establishment and popularization of the scientific method (Fig. 2.4). He is therefore considered to be one of the founders of modern science. His most important contribution was to articulate a new method of acquiring and testing knowledge of the world. Before him, many philosophers had come up with ideas about the nature of the world, but few of them had tested their ideas against reality. For example, the ancient Greek philosopher Thales postulated that everything in the world was made from water, but never tested his theory to establish whether it was

**Fig. 2.4** Francis Bacon
(Portrait by Paul van
Somer, 1617, Dulwich
Picture Gallery, Public
domain, via Wikimedia
Commons)

true or not. Bacon proposed that our scientific knowledge should be based entirely
on empirical evidence rather than on tradition, religion, or philosophical specula-
tion. He was one of the first to propose that scientists should start their studies by
meticulously collecting data from observations or experiments, and then carefully
analyzing and interpreting it. In addition, he stressed the importance of *induction* in
science. Induction is the process where general conclusions are drawn from a limited number of observations. For instance, if every swan you have seen in the past is
white, then you may hypothesize that all swans are white, or if the sun has risen
every day in the past, then you may hypothesize that it will always rise. Induction is
therefore important when developing theories and hypotheses to explain observations and to predict what might happen in new situations. These theories and hypotheses can then be tested and refined if needed by carrying out further observations or
experiments. Another key contribution of Bacon was his emphasis on the importance of skepticism – a good scientist should always question the assumptions
underlying their and other people's work – they should not rely on tradition or
authority. Finally, Bacon was a strong advocate of scientists sharing their findings
with each other and the general public because he believed this would advance our
understanding of the world more effectively, as well as lead to knowledge and technologies that would improve our quality of life.

Bacon was a major figure in the development of the modern scientific method
and his proposals are still important for practicing scientists today.

## David Hume

David Hume (1711–1776) was another key figure in the development of the scientific method (Fig. 2.5). He was part of the Scottish Enlightenment movement that
bloomed in the late eighteenth and early nineteenth centuries that included major

**Fig. 2.5** David Hume
(Portrait by Allan Ramsay,
1754, Public domain, via
Wikimedia Commons)

literary, philosophical, and scientific figures, such as the poet Robert Burns and the economist Adam Smith. One of Hume's major contributions to the scientific method was to question the philosophical basis of induction. He highlighted that just because every observation in the past has shown that something is the case, does not necessarily mean that it will always be the case. For instance, the fact that every swan seen in the past has been white does not necessarily mean that all swans are white. Indeed, black swans were found to exist in Australia. The fact that the sun has been seen to rise every day of human history, does not mean it will always rise. The sun may eventually explode. Hume pointed out that we are making some implicit assumptions about the way the world works when we use the inductive method. For example, we are assuming that the future will always resemble the past, which may not be the case. Thus, theories are always open to revision if new evidence arises. As discussed later, Karl Popper took this idea further and developed a new approach to testing scientific theories.

So, what can graduate students and other practicing scientists learn from Hume's critique of induction? It is important to be aware that there are assumptions underlying our theories and hypotheses. While these assumptions are often necessary to make scientific progress, we should be aware that they may not be valid. No theory can be proven to be absolutely true. We should, therefore, always be prepared to revise our theories and hypotheses in the light of new information.

## *Immanuel Kant*

Immanuel Kant (1724–1804) was a German philosopher who is considered to be one of the most influential thinkers of all time (Fig. 2.6). He made transformative contributions to many areas of philosophy, especially ethics (the theory of moral

**Fig. 2.6** Immanuel Kant
(Portrait by Johann
Gottlieb Becker 1768,
Public domain, via
Wikimedia Commons)

behavior), metaphysics (the theory of the nature of reality), and epistemology (the theory of knowledge). He was born and lived his entire life in the same small town in Prussia. His work is notoriously difficult to read and understand, which is partly due to the complexity of the issues he addressed, but also because of his obscure writing style. One of his most important contributions to the philosophy of science was the role of our senses and minds in understanding the world. He argued that our knowledge of the world is based on the way it is perceived in our minds, rather than on the real world itself, which is inaccessible to us. Information about the world must come from our sensory apparatus, such as our sight, hearing, taste, smell, and touch, before it is represented in our minds. Moreover, our minds are constructed in a particular fashion (which we now know is mainly due to evolution) that only allows us to represent and interpret our sensory perceptions in certain ways. For instance, our minds seem to be constructed to interpret the world in terms of time, space, causality, and continuity. Thus, we never have direct access to the world as it really exists, but only to the representation of it within our minds, which is limited by our sensory apparatus and the structure and function of our minds. We may perceive a tomato as a reddish ball-like object, but in reality, it is some kind of "matter" and "energy" that interacts with our senses in particular ways that lead to an impression of a tomato in our minds.

How is this insight of Kant useful to practicing scientists? An appreciation of the importance of our perceptual framework on our understanding of the world can help us to design and interpret our experiments better. The answers we obtain depend on the framework we adopt. Consequently, we should carefully examine our framework and identify any underlying biases within it – how does our conceptual approach and experimental methods influence the kind of information we obtain?

Kant was mainly concerned about the limitations of human perception. However, in science, our perception also depends on concepts, tools, and theories that augment our natural senses and minds like microscopes, telescopes, spectrometers, microphones, electronic noses, electronic tongues, and computers. For example, the first century philosopher Lucretius hypothesized that matter consists of tiny 'atoms' but he had no empirical means of confirming this. However, modern instrumentation allows us to prove the existence of atoms, with powerful techniques like scanning tunnelling microscopy even allowing scientists to directly view single atoms in real-time. The perceptual framework of scientists can therefore evolve as science and technology advance. Thus, our perception of the world now is different from that of past scientists and will continue to evolve. We should be aware that the information we obtain about the world does not only depend on what is actually out there, but also on how we acquire the information, which is influenced by the instruments, methods, and theories we use.

## *Karl Popper*

Karl Popper (1902–1994) was an Austro-British philosopher who spent his early life in Austria but emigrated to New Zealand when the Germans occupied it just before the start of the Second World war (Fig. 2.7). He then moved to England where he took up a position as a professor of logic and scientific method at the London School of Economics. During his life, he wrote numerous influential books and articles on society and the scientific method. His publications on the scientific method are thought to be some of the most insightful in the philosophy of science and have been very influential among practicing scientists.

Karl Popper applied David Hume's insights about the limits of induction in science. A theory can never be conclusively verified because we always have only limited access to the world – we do not have access to all time, space, and

**Fig. 2.7** Karl Popper in the 1980s (LSE library, No restrictions, via Wikimedia Commons)

information. It is always possible that we could find a case in the future that would not agree with our theory. Popper therefore rejected the classical notion that the purpose of the scientific method was to verify theories. Instead, he introduced the concept of *empirical falsification*. Although a theory can never be shown to be true, it can always be falsified by comparing its predictions with reality. If the predictions do not agree with the empirical evidence, then the theory is wrong and needs to be modified or replaced. Popper therefore argued that a theory was only scientific if it could make predictions about the world that could be falsified. He also stressed that theories are created by people using their imaginations to solve specific problems. These theories are always provisional attempts to understand, describe, and predict the world rather than being absolutely true. Theories therefore evolve over time as we test them against empirical data and find out where they are useful and where they need to be refined or changed.

As practicing scientists, what can we learn from Popper's analysis of the scientific method? When designing your research, you should clearly state your hypothesis in a way that it can be falsified. For instance, you could make the hypothesis that the level of vitamin D in people's bloodstream increases when they are exposed to sunlight. You could then assign people to two groups, one that is exposed to sunlight and one that is not. You can then measure the change in the vitamin D levels in their bloodstream before and after exposure. If the vitamin D level does not change, then your hypothesis is false. If the vitamin D level does increase, then your theory is provisionally "correct," but could be falsified in the future. Alternatively, if you are developing a theoretical model, you should design your experiments so that the model's predictions can be empirically compared to measurements or observations to determine whether the theory agrees with reality or not. As an example, you might have developed a mathematical equation that relates the hardness of a composite material to the concentration of particles it contains. You can then use the model to predict the change in hardness with particle concentration and carry out measurements to determine whether the theoretical predictions agree with the empirical data. If they do not agree, then the theory is not an accurate description of reality and needs to be further developed, or an entirely new approach may be required.

## *Thomas Kuhn*

Thomas Kuhn (1922–1996) was an American philosopher of science who had a major impact on our understanding of scientific practice and development (Fig. 2.8). He began his career as a physicist, earning undergraduate and graduate degrees from Harvard University, with some time in between serving as a radar specialist in the army during the Second World War. After his graduate studies, he became a faculty member at Harvard focusing on history and philosophy of science. While there, he published *The Structure of Scientific Revolutions* in 1962, which had a profound impact on the philosophy of science. Kuhn divided scientific progress into

**Fig. 2.8** Thomas Kuhn
(by Davi.trip, CC BY-SA
4.0 via Wikimedia
Commons)

two forms: "normal science" and "paradigm shifts." Normal science occurs in periods where most scientists working in a particular area adopt the same framework, involving specific well-accepted concepts and methods when performing their research activities. This leads to an incremental increase in our knowledge and understanding of the world. However, as existing hypotheses and theories are tested, there are more and more instances where they are found not to work. Scientists may then look for new approaches.

An example of a normal period of science according to Kuhn was the late nineteenth century, when most physicists used classical Newtonian mechanics as a framework to understand the world. This framework determined the kinds of questions that scientists asked and the problems they worked on, as well as the way they designed and interpreted their experiments. In contrast, "paradigm shifts" occur when one or more scientists come up with a new conceptual framework for carrying out science. This is often in response to the fact that the traditional theory did not describe certain aspects of the world well, so they were motivated to develop a new one. The concepts and methods of this new framework are typically incompatible with those of the traditional one. There may then be a period in science where some scientists use the conventional approach while others adopt the new approach, which causes conflict. During this period, people working with the conventional approach may be reluctant to change because they are so invested in it. Eventually, the new approach may replace the old one if it provides a better description of the world. Then, a new period of normal science begins as the majority of scientists adopt the new approach. An example of a paradigm shift in physics is when the theories of relatively and quantum mechanics were developed in the early twentieth century, which were incompatible with classical Newtonian mechanics. Eventually, most physicists adopted these new theories as they gave more accurate and comprehensive descriptions of the world. The adoption of this new framework then governed the types of scientific questions that were addressed, as well as the concepts and methods used to design and interpret experiments.

What can a practicing scientist learn from Kuhn's analysis of the scientific approach that will be useful for advancing their own research practice? We should always recognize that we may be strongly invested in a particular theoretical or conceptual framework because we may have worked with it for many years, and are therefore resistant to changing it. We should therefore always be mindful of the framework we are using and appreciate that it is contingent and open to revision. We should always keep an open mind to new developments in our field, and be willing to change if necessary. Even so, it is still important to rigorously and critically assess any new theories and concepts that form part of a paradigm shift, and only if they provide a better empirical description of the world, be willing to adopt them.

## Modern Theories

More recently, there have been some criticisms of the scientific methods proposed by Popper, Kuhn, and others. Philosophers, sociologists, and historians have shown that the way scientists actually go about their work is much more complex, diverse, and subjective than the well-defined and objective scientific method previously proposed. Scientists are people who have their own ambitions, temperaments, and goals. Although they may carry out their experiments with the intention of being objective, they often let their subjective motives creep into the design, performance, and interpretation of their studies. For instance, publishing numerous manuscripts in highly ranked academic journals increases the number of papers and citations they get, which can have significant career benefits for scientists such as enhancing prestige, improving promotion chances, increasing salaries, and strengthening funding opportunities. Consequently, there is a strong motivation to carry out experiments that lead to results that are impactful and publishable. This may lead to a lack of objectivity when interpreting the results of an experiment, such as ignoring data that does not fit your hypothesis or cherry-picking data that supports a new or interesting finding.

As a research scientist, it is important to be aware of your personal motivations and how they may impact the design, performance, and interpretation of your results. In your rush to get high impact results, you may lack objectivity, which reduces the value of your scientific contributions. Even though individual scientists may not always demonstrate the ideal of objectivity, the overall scientific enterprise is designed to overcome these prejudices. Science is both a cooperative and competitive endeavor. A person's work will be peer-reviewed, which helps to weed out some of the non-objective tendencies of an individual scientist (which may be conscious or unconscious). Moreover, the fact that scientific manuscripts are written in a way where other scientists can easily repeat your work means that people can test whether what you report is true. If it is not, then they can contradict you in the literature. Ideally, this process helps to eliminate any major errors in science.

We recommend that practicing scientists understand the framework they are working in. What kinds of questions does it address? What kinds of questions does it not address? (These may be areas for future research). What assumptions underlie it? Then, when working in the framework, always be aware that it is contingent – it is open to revision. Use the existing framework to make as much progress as possible but look out for areas where the concepts and theories do not fit reality. Highlight these in your published work. If possible, revise the existing theories or develop new ones that give a better description of reality.

## Pioneers and Settlers

Different kinds of scientists can be thought of as either pioneers or settlers striving to conquer a new land (Fig. 2.9). The pioneers adventure into unknown lands and create new frontiers. The settlers then move in and build towns, cities, farms, factories, reservoirs, communication networks, and transport systems. They may not have discovered the new lands, but they get to know the landscape better and make it more livable. Scientists who develop novel concepts or theories (like Galileo, Newton, Darwin, or Einstein) that open up new intellectual frontiers are pioneers. In contrast, scientists who adopt these ideas and put them into practice help to understand the landscape better and use the knowledge to create new technologies that make our world a better place to live. In this analogy, the scientists responsible for Kuhn's

**Fig. 2.9** Science is made up of pioneers and settlers. (*Image*: Pioneers of the West, Helen Lundeberg, Public domain, via Wikimedia Commons)

paradigm shifts are the pioneers, whereas those involved in Kuhn's normal science are the settlers. Both pioneers and settlers play critical roles in understanding and improving our world. In practice, most scientists are settlers, rather than pioneers.

## Transformative *versus* Incremental Science

When people think about scientific discoveries, they often think about the transformative ones made by the pioneers highlighted by Kuhn, such as the discovery of natural selection, relativity, quantum mechanics, and genes. The scientists involved in these paradigmatic shifts in our thinking are considered to be the rock stars of science. For this reason, many people strive to become pioneers who make these transformative discoveries. In contrast, scientists who carry out research that only leads to incremental increases in our knowledge are sometimes overlooked. However, many important advances in our understanding of the world, as well as many impactful technological innovations, are based on accumulated knowledge obtained from this incremental approach. Our understanding of how biological cells work has been based on small contributions made by scientists working in many different areas, including the development of microscopes to characterize the internal architecture of cells, the design of analytical instruments to characterize the structure and function of proteins, carbohydrates, lipids, nucleotides, and other biological molecules, and the development of systems biology approaches to understand complex integrated systems. The creation of modern smartphones relied on small advances in many areas, including miniaturization, integrated circuits, microprocessors, wireless communication, touchscreen technology, sensors, software, and batteries. Indeed, most transformative scientific discoveries are actually based on incremental knowledge that has been acquired over many years by numerous scientists.

Consequently, there is a strong argument for having a good balance between those scientists doing transformative work and those doing incremental work. Some people are pushing back the boundaries, whereas others are taking the new knowledge and putting it into practice so that it has a positive impact on our lives. Both types of science should be recognized as being important and celebrated. Many new graduate students dream that their work will be transformative and lead to major changes in the field. Whilst this is true for some, most scientists are actually involved in research that leads to more incremental advances.

## Reductionism and Holism

The approaches scientists adopt to understand our world can often be characterized as either reductionist or holist. Each of these approaches has its advantages and disadvantages, but both are important sources of knowledge. Moreover, a particular approach may be more suitable for a particular problem.

## *Reductionism*

Reductionism is an approach that seeks to understand complex systems by breaking them down into simpler, more fundamental constituents. It assumes that the properties of complex systems can be explained by understanding the nature, behavior, and interactions of their constituent components. For example, polymer chemists often assume that the mechanical properties of plastic packaging can be understood by studying the types of polymers they contain and how they interact with each other. Reductionism has been successful in many areas of science, especially physics and chemistry, where an understanding of the behavior of systems at the atomic and molecular levels provide valuable insights. This approach has led to many important scientific breakthroughs, including our understanding of the structure of matter, the laws of motion, and the nature of chemical reactions. Nevertheless, reductionism has limitations, especially for understanding highly complex and dynamic systems, like living organisms, ecosystems, or human behavior, because it cannot account for emergent properties that often arise in these systems. For example, it is currently impossible to account for the evolution of different species from only knowledge of the chemistry and physics of biological molecules. Nevertheless, reductionism is an extremely powerful approach in many areas of the physical sciences. In these disciplines, complex systems are broken down into simpler model systems whose properties can be understood using fundamental physical and chemical principles. Once the behavior of the simpler system is understood, its complexity can be incrementally increased until it more closely resembles the real system of interest.

Reductionism is therefore a valuable tool in scientific inquiry, which allows scientists to break down complex problems into manageable parts and to understand the underlying mechanisms. However, it is essential to also acknowledge its limitations and consider complementary approaches where necessary.

## *Holism*

Holism is an approach that emphasizes the study of systems as a whole, rather than focusing on the properties, behavior, and interactions of their constituents. It recognizes that a strictly reductionist approach cannot be used to understand the behavior of many complex systems, due to emergent properties that arise as a result of the complicated interactions between multitudes of different constituents. For instance, it is not possible to predict how people think by simply understanding the physics and chemistry of neurons. Thinking is an emergent property that arises from many complicated biochemical processes occurring inside the brain and other organs. Holism is particularly important in biology, ecology, and sociology for understanding living systems, ecosystems, and human behavior. This approach emphasizes the importance of examining the relationships, interactions, and synergies between the different constituents and how they contribute to the overall behavior of the entire system. Systems biology is an example of a holistic approach that is being increasingly used in biology to understand the behavior of living organisms and ecosystems.

Reductionism and holism are not mutually exclusive scientific approaches – they are often used in combination to obtain a better understanding of complex phenomena in the world. For instance, knowledge of the physics and chemistry of individual neurons can be used within a systems biology approach to develop a better model of the human brain. Similarly, understanding the inner workings of individual cells can help develop systems biology models of tissues and organs.

## General Approaches to Designing and Interpreting Research

Scientists adopt different approaches to carry out their experiments depending on the complexity of the systems they are studying and the type of results they are looking for. A few of these approaches are highlighted here, some of which overlap with the reductionist and holistic approaches discussed previously. An appreciation of the advantages and disadvantages of each of these approaches can help to select the most appropriate one to use when designing and interpreting experiments. It should be stressed that there is no single unified approach that can be used in all scientific disciplines to tackle all problems. Instead, scientists have to use their experience, creativity, and imagination to select the most appropriate one for their particular situation.

### *Trial-and-Error Approach*

The trial-and-error approach is commonly used in industry for solving manufacturing problems, as well as for developing new and improved products and processes. It may also be used in some academic laboratories that are developing new materials. In this approach, the investigator usually prepares a sample for study that has certain characteristics (*e.g.*, composition and structure) and then subjects it to some form of processing treatment (*e.g.*, heating, pressurization, or shearing). The properties of the resulting sample are then measured, and the investigator establishes whether the sample characteristics or processing treatment used can result in a product with the required properties. If the final product meets these requirements, then the initial sample characteristics and/or processing treatments are selected, but if it does not, then these parameters are changed and the procedure is repeated until a final product with suitable properties is obtained. A potential advantage of the trial-and-error method is that it's sometimes possible to rapidly solve a problem or develop a new product using minimal resources. For example, an investigator with some prior knowledge of a system may be able to rapidly select the optimum initial sample characteristics and processing treatments required to create a desirable final product. The major disadvantages of this approach are that it may not be possible to solve the problem in a reasonable time (if the wrong input values are selected), it may not produce a robust solution, or it may not produce the optimum solution to the problem. If a more rigorous study was carried out, it might have been possible to identify a solution that was more efficient, more robust, or less expensive. In addition, the trial-and-error method is largely dependent on the accumulated expertise of the investigator and provides little insight into the fundamental physical or chemical processes that govern the properties of the system being studied. This approach should therefore only be used when there is no alternative.

## Black-Box Approaches

The black-box approach is a somewhat more systematic approach of obtaining information about the system being studied, but it is also not concerned with providing information about the fundamental physical and chemical processes occurring within the system. Instead, the system (the "black-box") is subjected to one or more treatments (inputs) and the change in one or more properties of the system (outputs) in response to these treatments is measured (Fig. 2.10). The investigator then reports the measured system properties for each of the various treatments and/or uses statistical models to correlate the inputs to the outputs. To provide a concrete example of this approach, the system could be a plant-based milk, the treatment could be a change in temperature (input), and the measured property could be the viscosity of the end product (output). The investigator would prepare a plant-based milk, subject it to different temperature treatments, measure the viscosities of each of the samples, and then report the change in viscosity with temperature. This approach is particularly useful for identifying the major factors that influence the properties of materials, and in assessing their magnitude and relative importance. It is widely used as an initial screening procedure by investigators who are studying a system

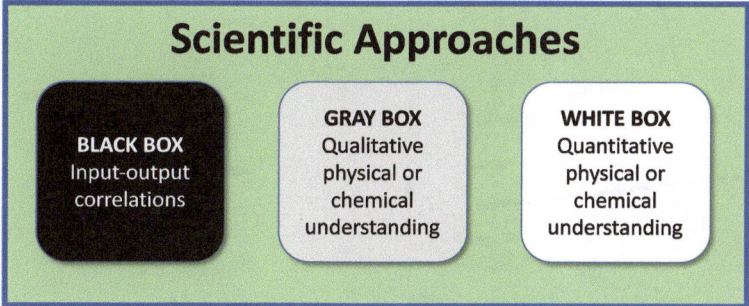

**Fig. 2.10** Different approaches can be used to understand the world using the scientific method, which vary in the insights they provide about the fundamental physical, chemical, biological, and other processes involved

that has not been extensively studied before since it enables them to rapidly develop an understanding of the dominant factors that influence its properties. In some situations, the black-box approach may be the only option available since the theoretical concepts or analytical techniques required to probe the internal operations of the system may not be available.

Nevertheless, the black-box approach has limited value when used improperly or taken to extremes. For example, experiments designed and analyzed based entirely on statistical models (such as surface response methodology) can produce results that are confusing or misleading. Rather than using existing knowledge of the fundamental physicochemical properties of the system to interpret the experiments, investigators select (often inappropriate) ranges of input and output variables based on some statistical model and then find statistical correlations between the input and output variables. For instance, surface response methodology may tell you to measure the strength of a protein gel at 0, 100, and 200 mM NaCl, assuming there is a smooth change in gel strength with salt concentration. But in reality, there may be a maximum in the gel strength between 0 and 100 mM NaCl that would be missed (Fig. 2.11).

Even so, statistical models can be useful in establishing the relative importance of different factors that may affect a system's properties, but they should always be used carefully and in combination with the white- or gray-box (physicochemical) approaches described later whenever possible. Indeed, if an investigator truly had no knowledge of the behavior of the system being studied, it would be difficult to select the type of input parameters to vary, the range of input parameters to select, and the type of material properties to measure. In practice, investigators usually have a fairly good *a priori* expectation of the factors that are likely to be important, which facilitates the selection of the most appropriate input and output variables to use.

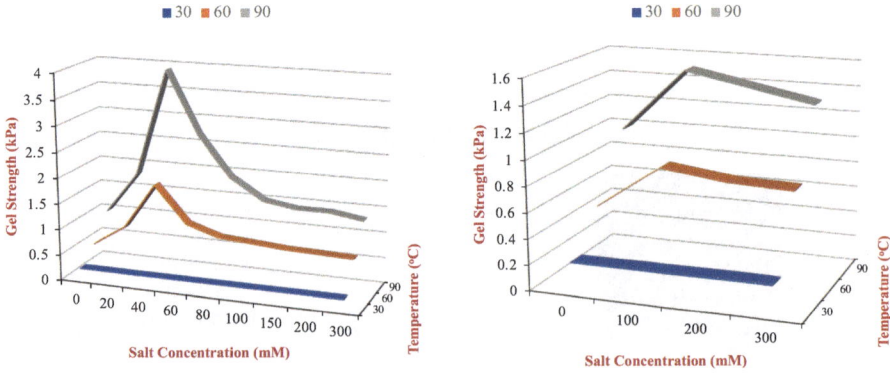

**Fig. 2.11** Using statistical methods to design experiments, such as surface response methology, can mean that important features are sometimes missed. Here, the impact of salt and temperature on the strength of protein gels is shown (left) but surface response methodology may miss out the peak because the range of salt concentrations selected is too wide (right)

Recently, there has been growing interest in the use of the black-box approach due to the rise in powerful AI software. This kind of software can help to establish links between different inputs and outputs, provided that enough experiments have been carried out to generate the high-quality data needed to train the program. The program can then be used to predict how a system will behave given certain new inputs. This approach can be very useful when optimizing the properties of systems. For instance, one of us (DJM) is working with a plant-based cheese company that is using AI to create products that closely resemble the desirable sensory attributes of real cheese, such as its look, feel, and taste.

## Gray- and White-Box Approaches (Physicochemical Approaches)

The main disadvantage of the trial-and-error and black-box approaches is that they do not provide any direct understanding of the fundamental physicochemical processes that occur within a system. Knowledge of these processes is usually desirable since it provides deeper and more quantitative insights into the factors that determine the overall system properties. Therefore, this significantly improves the identification of effective strategies for controlling system properties and can allow one to make predictions about how the system (or related systems) will behave under other conditions. For this reason, you should aim to use a fundamental "physicochemical approach" to understand the properties of the systems you are studying, whenever possible.

This approach attempts to understand why a particular system behaves the way it does when it is subjected to a particular treatment at a fundamental scientific level. Physicochemical approaches can be divided into "white-box" and "gray-box" approaches depending on the level of understanding one has of the system. In the white-box approach, the fundamental properties of the system can be quantitatively described using mathematical equations or computer simulations. For instance, the change in viscosity of a dilute suspension of small particles dispersed in water can be accurately described by Einstein's equation: $\eta = \eta_0(1 + 2.5\phi)$, where $\eta_0$ is the viscosity of pure water and $\phi$ is the volume fraction of the particles. This equation can be used to predict how the properties of a plant-based milk change when the concentration of fat droplets it contains increases. In the gray-box approach, the system in question may be so complex that its behavior cannot be quantitatively described, but physical or chemical principles can still be used to qualitatively interpret the observed trends. For example, the hardness of cheese may increase when salt is added, which could be attributed to the screening of electrostatic interactions between the protein molecules. In this case, it is difficult to accurately predict how the hardness will change when salt is added from first principles, but the data can still be interpreted in terms of known physicochemical phenomena.

Physicochemical approaches can be implemented in two broad ways depending on the starting point. First, an investigator could start with some empirical observation or measurement of a system and then try to establish its physicochemical origin using a range of analytical techniques, physical concepts, and mathematical models. For example, a researcher may find that a plant-based milk tends to aggregate and separate when added to coffee. They may then measure the pH of the coffee and the electrical charge on the fat droplets in the milk. As a result, they may find that the charge on the fat droplets is close to zero because the pH of the coffee is near the isoelectric point of the proteins coating the fat droplets. As a result, it can be deduced that aggregation occurs because the electrostatic repulsion between the fat droplets decreases when the plant-based milk is added to coffee. Second, an investigator could start with a conceptual or theoretical model and then use it to make predictions about how a real system should behave. For instance, they may have a theoretical model that predicts how the repulsive interactions between protein-coated oil droplets changes with pH, and how this effects the tendency for the droplets to agregagate. They could then use this model to prediect how adding these oil droplets to acidic coffee would impact their resistance to aggregation. These predictions could then be compared with experimental measurements made on a real system, and the investigator could determine how well the model describes the properties of the real system. If there are deviations between the theoretical predictions and experimental measurements, then the mathematical model could either be discarded or modified to take them into account. By comparing how closely theoretical predictions and experimental measurements agree, it is often possible to obtain quantitative insights into the physicochemical processes that determine the properties of systems. In reality, these two different ways of using the physicochemical approach are closely related to each other, and investigators often use a combination of them. The level of understanding that is achievable using the physicochemical approach is largely determined by the complexity of the system studied, as well as the sophistication of the analytical instrumentation and theoretical models available.

More recently, there has been an increase in the utilization of sophisticated computational models that are designed to simulate the behavior of physical, chemical, or biological systems and processes. These computational models are often used in drug design and material science. For instance, an imaginary box is created by the computer program that contains atoms or molecules of known properties (such as size, shape, charge, polarity, and flexibility) that represent those in the system being studied. The forces operating between these different atoms and molecules are then modeled by the software (such as van der Waals, electrostatic, and steric interactions). The computer program is then run and the change in the organization of the atoms and molecules over time is monitored. These programs can provide detailed insights into the behavior of materials at the molecular level (Fig. 2.12). This knowledge can help in understanding and controlling the properties of materials, which can lead to new and improved products, such as drugs, foods, cosmetics, or polymers.

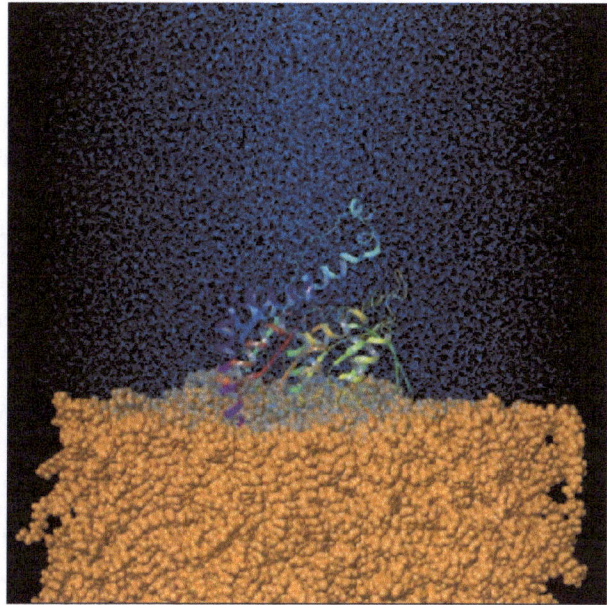

**Fig. 2.12** Example of the power of computer simulations for providing insights into the properties of complex systems. This image shows a potato protein adsorbed to an oil (orange) – water (blue) interface. Image kindly provided by Jeff Sanders (Schrodinger). This information could be used to create plant-based milks with improved stability

## Concluding Remarks

The scientific method has proven to be one of the most powerful tools humans have ever invented. It has radically transformed the world over the past few hundred years, helping to improve health, extend lifespans, and enhance the quality of life for billions of people around the globe. The scientific method requires researchers to focus on empirical data to prove or disprove their hypotheses, as well as to take an objective approach when designing, performing, and interpreting their experiments. An improved understanding of the scientific method and its limitations can help practicing scientists work more efficiently and effectively. Knowledge of the scientific method can help you to design and interpret your experiments. There are often several different approaches that can be adopted when carrying out a scientific study and it is important to identify the most appropriate one for your particular problem. The field of science is in a perpetual state of evolution and it is important to recognize that there may be more effective approaches to problem-solving than the ones presently employed. It is crucial to stay informed about the latest developments within your scientific discipline, as well as across other disciplines.

Maintaining an open mind and embracing change can contribute to the cultivation of strong scientific skills.

**Key Points**

- The scientific method can be divided into several parts: research question development, hypothesis generation, observation or experimentation, hypothesis testing, hypothesis reformulation, and theory development.
- Sources of information used by scientists include observations, surveys, experiments, theories, simulations, and search engines.
- A scientific study can be broken into three main phases. The creative phase involves coming up with new ideas, an original research question, and a method of tackling it. The research phase involves designing an experiment, performing trials, and analyzing the results. The communication phase entails disseminating the results in an effective manner.
- Many philosophers and sociologists have contributed to our understanding of the scientific method. Knowledge of their work is useful to research scientists.
- It is important to be objective as a scientist. There are many pressures to obtain certain results, and it is important to accurately represent the research performed (be aware of cherry-picking and other biases when interpreting results).
- Science can be transformative or incremental. Transformative science leads to pioneering discoveries, whereas incremental science leads to small advances in an established field. Transformative science is often given more attention and praise, but both categories are essential for scientific progress.
- Reductionism and holism are different approaches to understanding our world. Reductionism seeks to understand complex systems by breaking them into simpler systems. In contrast, holism examines complex systems in their entirety.
- Different approaches can be used to gain an understanding of scientific phenomena, including black-, grey-, and white-box approaches, which involve increasing levels of explanation in terms of fundamental scientific principles.

# Resources

Shapin S (1996) The scientific revolution. The University of Chicago Press, Chicago

Patten ML (2017) Understanding research methods: an overview of the essentials, 10th edn. Routledge, New York

Strevens N (2020) The knowledge machine: how irrationality created modern science. Liveright Publishing, New York

# Chapter 3
# Planning Your Research: A Laser Focus on Manuscripts

## Why Is Publishing Manuscripts So Important?

It is often assumed that scientific progress is the result of cutting-edge research. But it is really due to the publication of cutting-edge scientific manuscripts. You can perform great research, but if nobody ever learns about it, it has little value (unless you are working for a company that is protecting a trade secret). In this chapter, we highlight the critical importance of publishing in science. More than that, we stress that this is probably the *most* important thing a scientist does in their academic career. We also introduce an approach adopted in our labs that has proven to be extremely effective at increasing our research productivity. If you are a graduate student, adopting this method may help you complete your studies in good time, as well as to build a strong *curriculum vitae* that will advance your career.

Let's begin by considering some of the most important reasons for publishing your research in scientific journals:

- *Dissemination of scientific knowledge:* The results of your research cannot be utilized by other people if it is not readily available in the public domain. Once it is published, then other researchers can use it, thereby advancing our understanding of the world. Moreover, researchers and developers working in industry can use your work to create new products or improve existing ones, which helps translate your research findings into real-life applications. This is one of the most important and rewarding reasons for carrying out research – it helps to create new knowledge that improves the world we live in.
- *Advancing your career:* Publication of your research in peer-reviewed scientific journals provides concrete evidence of your creativity and productivity, as well as the relevance and impact of your work. If you want to become a professor, there are several metrics commonly used to rank and compare different researchers. The number of articles you have published in high-quality journals is important, as well as the number of citations of your articles by other researchers. These two metrics are combined in the Hirsh-index (h-index), which is the number of papers you have published that have been cited by other authors at least that same number of times. For example, if 45 of your papers have been cited at least 45 times, your h-index is 45. Typically, it is assumed that the higher your h-index, the stronger your scientific impact. Consequently, the more publications and citations you have, the stronger your *curriculum vitae* is, and the more attractive you become to future employers. Moreover, increasing your number of publications, citations, and h-index will help with future promotions, as well as make you more competitive for awards and scholarships.
- *Expanding your visibility and networking:* Publishing your research allows other scientists to know what you are doing, thereby increasing your visibility. This may have several benefits: it increases the possibility of collaborations with other researchers; it increases the chances of securing your next position, as a company or university may be looking for someone with expertise in your area; it increases the number of invitations to review scientific articles or present at scientific conferences, which enhances your *curriculum vitae*.
- *Beating the opposition:* Like many other areas of life, scientific research is both collaborative and competitive. In some situations, it is advantageous to work in multidisciplinary teams with other scientists to tackle complex problems that require a range of skill sets to solve. In other situations, different research groups may be in intense competition with each other. Competition is especially fierce when scientists are working at the cutting edge of research. In this case, it may

be extremely important to finish your research and publish it before a competitor can. A recent example of this was the discovery of the precise gene editing tool known as CRISPR Cas-9. Several groups were working in this area at the same time and were close to publishing their research findings. As a result, there was strong pressure to be the first to publish and gain the main credit for this ground-breaking technology. Profs. Emmanuelle Charpentier and Jennifer Doudna were the first to publish their manuscript, which meant that they got the most recognition and credit, eventually leading to them being awarded the Nobel Prize in Chemistry in 2020.

- *Job satisfaction:* Another important reason for publishing your research is that it is often highly rewarding. It can be really exciting to see your first paper published in a scientific journal. It is concrete evidence of all the hard work you have put into designing and performing experiments, analyzing the data, and writing the manuscript. It also proves that you can perform work that meets the standards set by the academic world.

- *Completing your thesis:* One of the most important reasons for publishing your research is that it will help you complete your thesis in a timely manner. Many institutions expect you to have published one or more manuscripts during your PhD. If your work has been published in a reputable journal, then it has already been peer-reviewed and accepted by the wider scientific community, which means that it is more difficult for the committee members on your thesis to reject it. Moreover, your research will already be prepared in a well-organized and well-written fashion, which will make it easier to write your thesis. Indeed, it may simply be necessary to reformat your manuscript so that it conforms with the thesis requirements of your institution. If you have published several articles, then these may serve as different chapters in your PhD thesis, so you have much less work to do. As well as carrying out original research, it is usually advisable for students to write a review article towards the beginning of their graduate studies. This allows them to get up to speed with current knowledge in the field and can be used as the basis for the literature review chapter in the thesis.

## When Do you Start Writing Your Manuscript?

Traditionally, people often start to write a manuscript after they have carried out all their experiments and analysis (Fig. 3.1). However, from our experience, the best time to start writing a manuscript is before you have carried out any experiments or only a few preliminary experiments. These preliminary experiments are mainly used to establish whether the materials and methods you intend to use lead to the kind of results you are expecting (Chap. 4). Designing the paper first might seem strange because you have little or no results to write about yet. But we find this to be an extremely efficient and effective way of planning and performing research. Adopting this approach typically makes your research much more focused, meaning that you will need to perform far fewer experiments in the long run. It should be

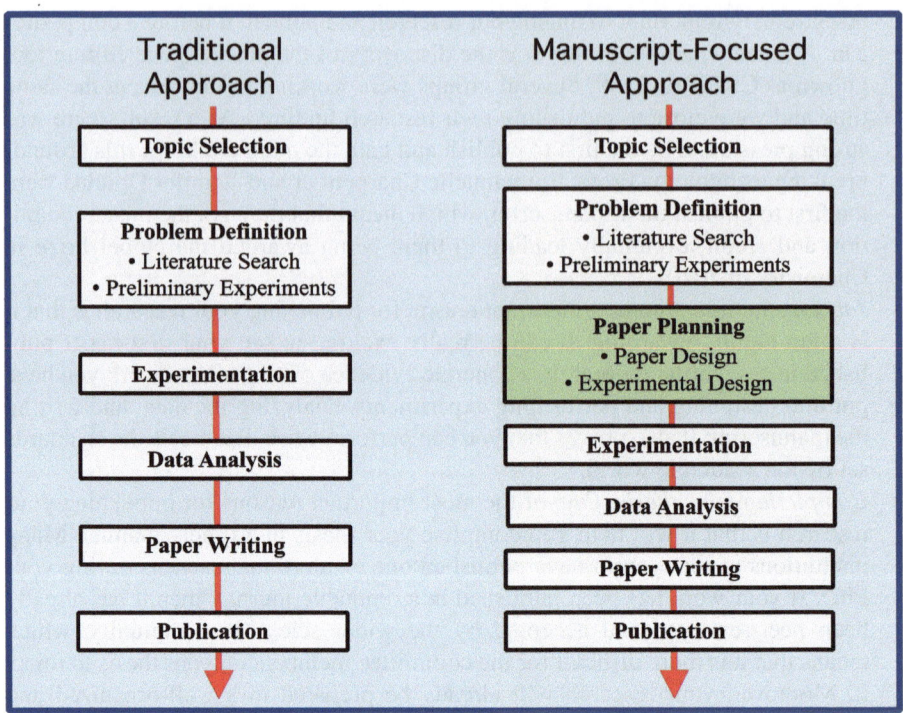

**Fig. 3.1** Comparison of the traditional and manuscript-focused approaches of designing and carrying out research that leads to a scientific manuscript

stressed that at the beginning of your research career, this approach is more difficult to implement because you do not have all the knowledge required. In this case, you can work with your supervisor to decide on the most appropriate experiments to perform. However, as you become more familiar with your research field, including the methods and concepts involved, you will be able to use this approach more independently. For this reason, we often encourage our students to start their research by writing a review article. This means they have to search the scientific literature to identify all the studies that have previously been carried out in their research area. They must critically evaluate these studies and organize them into a coherent story. This helps them to rapidly become an expert in the field, including establishing the current state of knowledge in their area, and where any controversies or gaps exist. It also provides them with insights into the nature of the experimental methods and theories typically used in the field. This knowledge is invaluable for designing their own original research papers. Moreover, the review article can be used as one of the opening chapters in their thesis.

When one of our graduate students is beginning a new project, we first meet to discuss the intended research. Before starting, we usually already have a good idea about what the general research area is. For instance, it might be using nanotechnology to develop a new encapsulation technology that increases the bioavailability of nutraceuticals, or it might be using polymer science concepts to create better quality plant-based meat, fish, or egg analogs. But we still need to decide on a specific topic that the proposed research will focus on. We find the best way to do this is to imagine what a scientific manuscript would look like if the research was successful. We therefore ask our students to complete a worksheet that gets them to think about the nature of the final manuscript, which is summarized in Table 3.1.

This worksheet contains various elements that must be considered when designing a research project that is suitable for publication. We provide some specific examples of this approach using our own research areas:

- *Title*: What is the working title of your manuscript? Coming up with a good title helps you to narrow the scope of the research and to focus on a particular problem. As an example, I may come up with a title "Utilization of edible nanoemulsions to increase the bioavailability of curcumin in functional foods: An *in vitro – in vivo* correlation." This may be changed after the research has been completed, but it provides a good focus to start.
- *Objective*: What is the main objective of your research? You should write a sentence or two stating the specific purpose of your proposed research. For our example, this could be: "The objective of this research is to use *in vitro* digestion studies and *in vivo* human feeding studies to determine the influence of encapsulation of curcumin in nanoemulsions on its bioavailability."
- *Relevance*: What is the relevance of your research? You should write a sentence or two stating the reasons why the proposed research is important. For instance, what new knowledge will be created? What gaps in the existing knowledge are being addressed? What problem is being solved? Doing this will help you to establish if the results you obtain are suitable for a scientific publication, assuming that the experiments went as planned. For our example, this could be: "Curcumin is a chemically unstable molecule with a low water-solubility and poor oral bioavailability. Encapsulating it within nanoemulsions may increase its

**Table 3.1** Worksheet for implementing the manuscript-focused research design approach, which can enhance research efficiency

| Paper element | Description |
|---|---|
| Title | Write the title of your paper |
| Objective | Write a sentence or two on the main objectives of your paper |
| Relevance | Write a sentence or two why this research is important |
| Novelty | Write a sentence or two on the novelty of your manuscript |
| Sections | Decide on the subject matter of each sub-section in the *Results and Discussion* section of the manuscript (typically 3–5) |
| Figures | Sketch the figures you expect to show in each section, assuming that all your experiments go as planned |
| Materials & Methods | Decide what types of materials and methods you require to obtain the data based on the figures you have sketched. |

The worksheet should be completed *before* carrying out the majority of your experiments or observations

dispersibility, stability, and efficacy, thereby improving its application as a health-promoting ingredient in foods."

- *Novelty*: What is the novelty of your research? Original research articles should provide new knowledge or insights that expand our current understanding of the field. Typically, you do not want to repeat research somebody else has published before, unless it was an important and/or controversial study, and you are trying to verify the results. You will therefore need to do a comprehensive literature search to find out what has been done previously, what is the current state of knowledge, and where any gaps in understanding or controversies exist. You should then write one or two sentences highlighting the novelty of your proposed research. For our example this could be: "Previous researchers have shown that nanoemulsions can be used to improve the water-dispersibility, chemical stability, and bioaccessibility of curcumin using *in vitro* studies, but there have been no previous studies relating *in vitro* digestion to *in vivo* human feeding studies."

- *Hypothesis*: After you have developed a good idea for a paper you should be able to formulate a hypothesis. A scientific hypothesis is a testable statement or proposed explanation for some phenomenon that can be observed. It provides a clear and concise summation of a scientist's prediction of what will be found in a study. It should be framed in such a way that it can either be accepted or rejected based on the results of the study. For example, a hypothesis might be: "We hypothesize that an increase in the bioaccessibility of curcumin observed in *in vitro* digestion studies will be positively correlated with an increase in the oral bioavailability of curcumin observed in *in vivo* human feeding studies." If the experiments show that an increase in bioaccessibility leads to a statistically significant increase in bioavailability, then the hypothesis is accepted but if they do not, then the hypothesis is rejected.

Once you have established that the subject of your manuscript is novel, important, and relevant, and you have formulated a strong hypothesis, you can then start designing your manuscript and experiments. We find the following approach to be useful:

- *Results and Discussion sub-sections*: Your final manuscript will include a *Results and Discussion* section where you present your findings and interpret them (Chap. 6). This typically involves providing molecular, physicochemical, and/or biological interpretations of your results, comparing your results to those reported by other researchers, discussing why your results are important, and highlighting how they advance the current understanding of the field. The *Results and Discussion* section is typically divided into several sub-sections (often around 3–5) that focus on different aspects of the research. Before doing any experiments, it is often a good idea to decide what these different sub-sections might be. For instance, the sub-sections for our proposed paper entitled "Utilization of edible nanoemulsions to increase the bioavailability of curcumin in functional foods: An *in vitro* – *in vivo* correlation" might be: (i) Nanoemulsion fabrication and characterization; (ii) Impact of environmental stresses (pH, ionic strength, and temperature) on nanoemulsion stability; (iii) *In vitro* gastrointestinal digestion of curcumin-loaded nanoemulsions; (iv) *In vivo* study of the bioavailability of curcumin-loaded nanoemulsions using a human feeding study (Fig. 3.2).

- *Figures and Tables*: After specifying the different sub-sections that might appear in the *Results and Discussion* section of your final manuscript, we encourage you to think about what kind of data will be generated during the research, and what the figures and tables might look like. For each sub-section, you should sketch several figures (typically 3–4) that are used to present the anticipated data in the final manuscript. You should draw the x- and y-axis titles and think about what kind of trends you expect in the data based on your current knowledge. Of course, there is no way of knowing whether these will turn out to be correct – this can only be established by carrying out the experiments. For instance, in the "Nanoemulsion fabrication and characterization" sub-section of the proposed manuscript discussed earlier, we may have several figures: (a) the particle size distribution of the nanoemulsions (particle concentration *versus* particle diameter), (b) the morphology of the nanoemulsions (transmission electron microscopy images), and (c) the impact of homogenization conditions on nanoparticle formation (particle size *versus* homogenization pressure) (Fig. 3.3). The same procedure is performed for each of the different sub-sections in the *Results and Discussion* section. At the end of this procedure, you will have a good idea of the figures and tables that might be present in your final manuscript. This can help you to design your research, as well as to assess whether the proposed research would lead to a scientific manuscript that is suitable for publication. You may decide at this stage that the manuscript would represent an important advancement in the field, or you may decide that the objectives need to be revised to make it more suitable for publication. This can save a lot of time in the long run.

- **Paper Title**
  - Utilization of edible nanoemulsions to increase the bioavailability of curcumin in functional foods: An *in vitro – in vivo* correlation
- **Results and Discussion Subsection Design:**
  - *Section (i)*. Nanoemulsion fabrication and characterization;
  - *Section (ii)*. Impact of environmental stresses (pH, ionic strength, and temperature) on nanoemulsion stability;
  - *Section (iii):* In vitro gastrointestinal digestion of curcumin-loaded nanoemulsions;
  - *Section (iv): In vivo* study of the bioavailability of curcumin-loaded nanoemulsions using a human feeding study.

**Fig. 3.2** It is recommended to make a list of the different sub-sections that will be used in the *Results and Discussion* section of a scientific manuscript before designing and starting your experiments

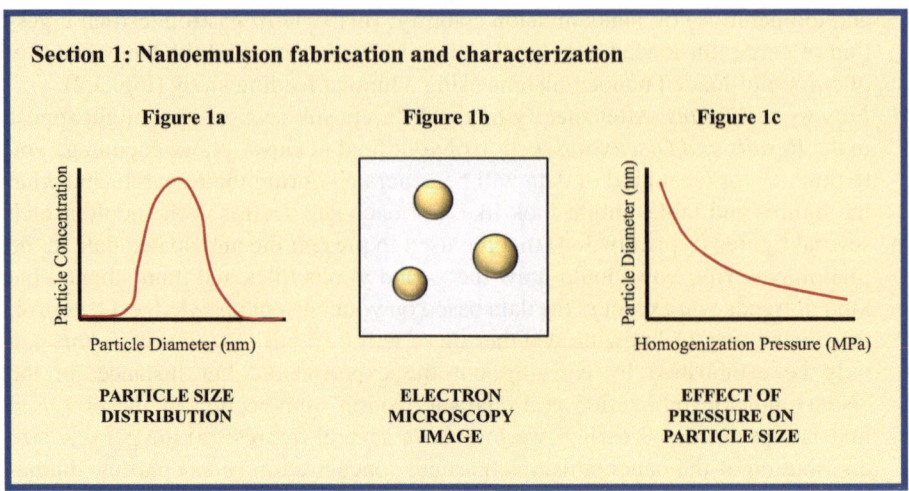

**Fig. 3.3** It is recommended to draw the figures that are expected in each section and subsection of the manuscript before starting any experiments

- *Materials and Methods*: Finally, you should have a good idea of what kind of research you need to perform to obtain a good scientific manuscript. At this stage, you should be able to specify all the materials (reagents, solvents, glassware *etc.*) and methods (instrumentation and protocols) that you need to perform the proposed research. You can then make sure all the relevant materials and methods are available, and that you learn any new experimental protocols needed.

Talking to our colleagues, we find that this manuscript-focused approach is not commonly used. Instead, graduate students are typically given a particular problem to work on where they have to define the problem, carry out experiments to address the problem, analyze the resulting data, and then design and write the paper (Fig. 3.1). This traditional approach often involves performing many more experiments than the manuscript-focused one, leading to a lot of data that is not used in the final publication. In addition, it can mean that some key experiments are not carried out during the initial experimental stage because the researcher was not focused on the overall picture of the study. As a result, a researcher may have to carry out additional experiments after they start writing the manuscript because some important information is missing. On the other hand, this approach may lead to new discoveries that might not have been found by taking a more manuscript-focused approach. Moreover, the manuscript-focused approach is not suitable for all kinds of science. It is more appropriate for the applied sciences where the expected results of an experiment can be predicted quite well, rather than for the basic sciences where they cannot. Even so, adopting a manuscript-focused approach may still benefit researchers carrying out fundamental research because it helps to clarify what kind of research is novel and relevant, as well as to establish the experimental techniques and methods required to address the problem.

## Be Open-Minded and Willing to Change

In the previous section, we highlighted a "manuscript-focused" approach that we have found to be highly effective for designing experiments that lead to scientific publications. It should be stressed that this approach only provides guidance for planning your research – your actual results may turn out to be very different than you expected. You must be objective when performing and interpreting your research. You should always have an open mind and look out for results that do not meet your expectations. This is because some of the most important scientific advances are made when the world turns out to be different than we first expected. You should design, perform, analyze, and interpret your experiments as carefully as possible, and if any results arise that do not meet your expectations, you should systematically try to find out why. It may be that there was a flaw in your experimental design or data analysis methods. It may be that you made a mistake in performing your experiments. It may be that you used the wrong chemicals or reagents, or they were passed their use-by date. It may be that the analytical instrument you used was broken or not calibrated properly. You must therefore think rigorously about all the potential things that could have gone wrong with your experiment. Creating a checklist can be useful for doing this. If you do this, and you still find unexpected results, then you may have made an exciting new discovery. Ideally, you should then look for a plausible scientific interpretation of your new findings.

## Types of Scientific Publications and Their Importance

There are several different formats available to publish your scientific findings, and it is important to be aware of their relative advantages and disadvantages. We therefore highlight the most common types of scientific publication here:

- *Original research articles*: These are usually the most important scientific manuscripts. These manuscripts are published in academic journals and report the findings of research studies on a specific topic. They are designed to provide new knowledge and insights into particular subjects by carrying out original research, which may involve experiments carried out in a laboratory, observations of the natural world, computer simulations, or population surveys. The manuscripts based on this work include descriptions of the importance and relevance of the research, the materials and methods used, and the results obtained, as well an interpretation and contextualization of these results. They often end by highlighting what new knowledge has been obtained and where future work is required.
- *Review articles*: Review articles also play an important role in science, and reading and writing them can have a major impact on your success as a scientist. This type of manuscript typically provides a critical overview of a particular research topic and is usually written by an expert in the field. For instance, a review article may be entitled "Utilization of nanoemulsions as delivery systems for nutraceu-

ticals." This manuscript might critically assess all the published research on the use of nanoemulsions to improve the dispersibility, stability, and bioavailability of nutraceuticals. When you start working in a new research area, it is important to carry out a thorough review of the literature to identify any good review articles published by other researchers. Reading these articles can quickly get you up to speed on the current status of knowledge and help to identify where gaps and controversies exist. It can also help to establish what kinds of experimental techniques are commonly used in your research area, which can be very useful when designing your own studies. Once you have carried out a comprehensive survey of the literature, you may be in a position to write your own review article. This is often recommended because it forces you to gain a detailed and thorough understanding of your field of interest. If you are a graduate student, then another benefit of publishing a review article is that it can be used as one of the first chapters in your thesis, which makes it much easier to write when you are nearing the completion of your degree. It also helps to establish you as an expert in the field. Often, review articles are cited more frequently than original research articles, which can be good when building your curriculum vitae, since it will increase the total number of citations you have. However, writing original research articles is preferred and should be a priority.

- *Brief communications:* These are usually short articles, sometimes known as *short reports* or *letters*, that an editor believes are of interest to the readers of the journal, and which are intended to stimulate further research or discussion in a specific area. They may also be used to publish new data in a rapidly emerging field so that a researcher can establish their claim as the first to make an innovative discovery. These brief communications may then be followed up by a more comprehensive original research article once more data is available. There are often strict word/figure/table limits for these kinds of articles.
- *Book chapters*: The purpose and format of book chapters are usually similar to those of review articles, but sometimes they can be more like original research articles, depending on the nature of the book. There are advantages and disadvantages of publishing your work in book chapters. The main advantages are that your chapter may be grouped with other chapters on a similar topic (which increase its visibility), you see your name in a book (which can be rewarding), it increases your visibility, and you may get a free copy of the book (which is always nice). The main disadvantage is that book chapters are often less accessible than journal articles, and so they have less impact and get fewer citations. Moreover, the research published in book chapters is often of poorer quality than that published in academic journals because it does not usually undergo the same level of rigorous peer review. Whenever possible it is therefore recommended for researchers to prioritize publishing their work in academic journals rather than as book chapters.
- *Encyclopedia entries*: Occasionally, researchers get asked to write encyclopedia entries on specific topics, which are usually related to their area of expertise. This kind of publication is increasing in popularity with some open access journals. Encyclopedia entries are often relatively short articles that focus on a particular

subject (*e.g.,* "nanoemulsions", "nutraceuticals", or "polymers"). They are typically not read or cited as much as articles published in academic journals. It is therefore recommended to publish your original work in an academic journal rather than in an encyclopedia. However, if it is relatively simple and quick to write, then it may be a good idea to contribute to an encyclopedia.

- *Conference abstracts and papers*: If you present a talk or a poster at a scientific conference, the organizers may ask you to submit an abstract of your research. This may then appear in the on-line and/or printed versions of the conference proceedings. In some cases, the proceedings of a scientific meeting may be published in a special issue of an academic journal or in the form of a book chapter. If it is published in a journal, it may have the same visibility and standing as an original research article. But if it is published as a book chapter, it may not get much visibility. Consequently, you should be careful not to publish your most impactful work in a format that might not be read much.
- *Trade articles*: Some scientific disciplines have trade journals that are read widely by both academics and industry researchers. These journals are often looking for interesting articles to publish and may ask professors, postdocs, or students to write a manuscript based on their area of expertise. These manuscripts may increase your visibility in the field, but they are typically not as highly cited as manuscripts published in academic journals. Again, it is therefore recommended to publish your research findings in academic journals rather than as trade articles.

## Where to Publish Your Manuscript

Once you have written a scientific manuscript and decided you are going to publish it in an academic journal, then it is important to select the most appropriate one. There are thousands of academic journals available, and you must decide the one that is most suitable for your manuscript. Some of the most important factors to consider are the subject matter, manuscript type (original research or review), journal quality (impact factor), publisher reputation (predatory or not), publication model (subscription or open access), and cost (free or paid). These factors are dealt with in more detail in Chapter 7, when we consider submission and revision of scientific manuscripts.

## Citations: Important (But Not Always)

An important parameter used to judge the success of a scientific manuscript, and the researchers who wrote it, is the number of citations it receives from other people. Typically, the greater the number, the more impactful the paper. Consequently, authors often aim to increase the number of citations of their manuscripts. There are several ways of doing this:

- Publish your work in influential journals that are well recognized in the field. Typically, these journals should have a large readership and high impact factor (see Chap. 7).
- Publish your work in open access journals, rather than subscription journals, as this increases the number of people who have access to your manuscript.
- Publish your work in academic journals rather than book chapters or trade journals, since the former are easier to find in literature searches and more accessible.
- Publish review articles, since they tend to get cited more than original research articles.
- Promote your work after it has been published, for example by sending it to your press office (if your findings have general interest) or posting a link to your manuscript on a suitable online forum (such as LinkedIn or X).

It should be stressed that the number of citations does not always reflect the quality or importance of a scientific manuscript. A manuscript can receive few citations for several reasons. It could on a topic that few other people are working on because it is irrelevant or trivial. It could be because the topic is in an emerging area that does not get many citations now but may in the future. It could be because the manuscript is on a topic that a lot of other people are already working on, and so it is just one of many. Hence, when someone wants to cite a manuscript on this topic, they have lots of others to choose from. It could be because a manuscript is published in a journal that is not included in online search engines. Finally, it could be because a manuscript is on a topic that is so complicated that very few other people have the skills to work on it (like some areas of mathematics or physics).

## Reasons for Delaying or Not Publishing

So far, we have stressed the importance of writing and publishing scientific manuscripts. Typically, you want to write and submit your manuscripts as soon as you can. However, there are situations where the publication of a scientific study may be delayed or even prevented. This can be extremely frustrating if you are a graduate student or postdoc since some (or all) of your hard work may not be published or takes a long time before it sees the light of day. This could occur if your supervisor is applying for a patent based on your research. As a result, you must delay the submission of your manuscript until the patent has been written and filed with the patent office. On the other hand, you may get your name as a co-inventor of the patent, which would strengthen your scientific reputation and *curriculum vitae*. Another reason publication may be delayed or prevented is because your research is funded by a company, which has an agreement with your university that the results of a specific research project are fully or partly owned by the company. For example, they may think that the research you are doing is highly innovative and could lead to new or better commercial products or processes. Consequently, they may want to delay or stop publication so they can use your results to file their own patent or trade secret. Finally, publication can be delayed or prevented if you use specialized ingredients or other substances in your research that have been provided by a company and are covered by a *material transfer agreement*. For example, a company may provide you with a specialized chemical to use in your research that is unavailable from other sources. The company who provided you with this chemical may ask you to sign an agreement to protect their intellectual property and trade secrets. Sometimes, these agreements specify that you can only publish the research if they agree, or that you must send the manuscript to them before you submit it to an academic journal so they have time to review and comment on it. It is therefore important to check with your professor if you are using an ingredient/component covered by a material transfer agreement.

## Concluding Remarks

Publishing manuscripts is the main way that scientists communicate their findings to other scientists, as well as to policymakers, companies, the media, and the general public. It also helps scientists to build their reputation, thereby increasing their standing within their field of expertise. If you are a graduate student, then designing, writing, and submitting manuscripts should be one of your main priorities. Publishing manuscripts during your studies makes it much easier to write your final thesis. It is strongly recommended to write a review article on your research topic that can be used as one of the introductory chapters in your thesis. It is also recommended to write several original research articles that can be used as subsequent chapters. This is an ambitious goal and depends on the nature of your research. It is more realistic for graduate students working in the applied sciences than those in the basic sciences. In this chapter, we presented a manuscript-focused approach for designing research projects, which we have found to be extremely effective in our own work. In this approach, a checklist is used that asks the researcher to provide the title, objective, relevance, novelty, and hypothesis of their proposed manuscript before starting most of the experiments. Moreover, the checklist asks the researcher to design the different sections and subsections of the manuscript, as well as to draw the expected figures and tables that will appear in the final article. This approach helps you to establish whether the manuscript would be suitable for publication if all the experiments went to plan. If you are a graduate student or postdoc, you can then show the checklist to your supervisor and get their feedback before starting your research. Of course, the results of scientific research are often unexpected, especially if you are working in a cutting-edge area. Consequently, it is always important to be objective and have an open mind that does not ignore unexpected results – as these may be the most interesting and impactful ones.

In conclusion, writing scientific manuscripts is a critical part of research, and can help you to build a strong reputation in your field of study, which may help you secure promotion or your next position. You should therefore prioritize publishing strong scientific manuscripts in well-respected academic journals throughout your research career.

**Key Points**
- The publication of scientific research is critical to the advancement of science as it lets others know what you have done, including other scientists, policymakers, the media, and general public.
- Publishing your research is also important for establishing your reputation and for career advancement.
- Writing a review article at the start of your graduate studies gives a comprehensive understanding of your research field, which helps guide your research and ensure it is novel. A review article can also be used as the first chapter of your thesis.
- An effective way of ensuring you are performing publishable work is to plan a paper before you begin research. You should consider the title, objective, relevance, and novelty of your work. You can then design results and discussion sub-sections, which will help to guide your experimentation.
- It is important to be open-minded and look out for results that do not conform to your expectations. Be willing to adapt your study as new results are found.
- There are many types of scientific manuscripts with different functions, including original research articles, review articles, book chapters, conference abstracts, and trade articles.
- It is important to select an appropriate journal to publish your work.
- The number of citations of your article is an indicator of its impact. Citation numbers can be increased by publishing in influential or open access journals. Review articles often gain more citations than original research articles but are often less prestigious.
- Typically, you would want to publish your work as quickly as possible but publication may be delayed in some circumstances, such as patent applications or funding from a commercial company.

## Resources

Belcher WL (2019) Writing your journal article in twelve weeks, second edition: a guide to academic publishing success. The University of Chicago Press, Chicago

Gastel B, Day RA (2022) How to write and publish a scientific paper, 9th edn. Greenwood, Santa Barbara

Heard SB (2022) The scientist's guide to writing: how to write more easily and effectively throughout your scientific Career. Princeton University Press, Princeton

Olson R (2015) Houston, we have a narrative. Chicago Press, Chicago

Schimel J (2012) Writing science: how to write papers that get cited and proposals that get funded. Oxford University Press, Oxford
Springer (2023) Writing a journal manuscript. Springer Scientifc, New York
Turabian KL (2018) Manual for writers of research papers, theses, and dissertations, 9th edn. The University of Chicago Press, Chicago

# Chapter 4
# Research Practice: Efficiently and Effectively Performing Your Experiments

## Introduction

In this chapter, we discuss the most important practical factors to consider when carrying out your experiments. Our main focus is on giving you advice that will streamline the process and make your research more efficient. In general, a rigorous scientific study can be divided into five main steps:

1. *Problem identification and definition.* In the first step, you identify a suitable problem to work on and clearly define your research question. This typically involves an extensive literature review to determine what is currently known and unknown about the problem.
2. *Experimental design.* In the second step, you define the parameters that will be measured or observed in the study, as well as the materials, protocols, and techniques that will be used to perform the study. As part of the experimental design, you will have to decide how many measurements or observations you intend to make in each part of the study.
3. *Data collection.* In the third step, you carry out the experiments, collect the data, and store them.
4. *Data analysis and interpretation.* In the fourth step, you collate, analyze, and interpret the data you have collected in your study.
5. *Communication.* In the fifth step, you prepare the results of your study for publication in scientific journals, submission of a patent, or presentation at a scientific meeting.

In the previous chapter, we introduced a manuscript-focused approach for efficiently identifying and defining a research problem (Step 1), as well as designing the experiments to be carried out (Step 2). This approach helps to systematically identify the most suitable chemical reagents, equipment, protocols, and analysis methods needed to perform the study. It also helps to ensure that if the research is successful, it will lead to new results that are innovative, relevant, and publishable. In this chapter, we focus on some of the more practical things to consider when designing and performing your scientific studies (Steps 2 and 3). Real-life examples from our own research experiences are used throughout this chapter to illustrate these best practices. In the following chapter, methods of analyzing and presenting your data are covered (Step 4). The effective communication of the results obtained through scientific publications and presentations (Step 5) is discussed in later chapters.

## Be Prepared

Before you start your experiments, it is important that you have everything you need to successfully carry out the study. You do not want to embark on your research project and then find out when you are halfway through that you do not have a critical chemical or the equipment you need is broken or unavailable. In the case of the physical sciences, it is important to ensure you have all the chemicals, reagents, equipment, and lab spaces needed to perform your study. You should also be familiar with all the experimental protocols needed to carry out the research, such as how to prepare and characterize your samples. In some cases, it may be advantageous to order the chemicals needed only after the preliminary experiments are completed, especially if these chemicals are expensive or have a short shelf life.

It is often a good idea to make a list of the reagents you need and then check to see if they are already available in your laboratory or department. If not, then you will have to purchase them. Chemical reagents can be expensive and can take some time to be delivered. Consequently, it is important to consider these factors when planning your experiments. It is important to ensure that the chemical reagents used are the ones you actually need for your study because the same reagent is sometimes available in different forms, such as different isomers, purities, or molecular weights (for polymers). Moreover, some reagents, especially biological reagents, degrade during storage and have a use-by date. For this reason, you want to be sure that all your reagents are fresh before using them. Another issue is that the composition and properties of some chemical reagents vary from batch to batch. Consequently, your results may vary if you change one or more of the reagents midway through your study. It may therefore be important to ensure you have enough of all the reagents to complete the intended study. This will mean that you have to calculate the quantities of the reagents needed to complete all the experiments you propose to carry out. Moreover, you should try to identify the source of any variations in the reagents and to characterize and report the composition and properties of the reagents used in your study. If a chemical reagent is too expensive, you may have insufficient funds to purchase it for your experiments. You may then have to redesign your study to use a different reagent.

It is also important to ensure you have access to all the equipment needed for your experiments, which might include simple things such as pH meters, balances, burettes, or mixers, or more specialized equipment such as chromatography, electrophoresis, spectroscopy, calorimetry, imaging, or microscopy instruments. The equipment should be available, operating properly, and clean. Many laboratories have numerous people using the same equipment. Consequently, you may have to reserve a time when you can use the instrument. Moreover, many instruments become dirty over time or contaminated with other people's samples, so it is important to clean them properly before using them; otherwise, your results may be compromised. The operation of many analytical instruments, including pH meters, balances, and chromatograms, changes over time, so it is important to calibrate them prior to use. Moreover, you may need to get trained on how to operate the equipment properly. Many modern scientific instruments are very easy to operate, but they can give incorrect results if not used appropriately. It is recommended that you learn the principles behind any critical piece of scientific equipment used in your research, as this helps to prevent you from using it incorrectly and generating results that are incorrect. If you are a graduate student, you may need to explain the principles behind any instrumentation used in your research in your graduate thesis and to your professors during your final defense. Finally, you may need to confirm that all the workspaces you need in your study are available and suitable for your research, *e.g.,* they may need to have fume hoods, glove boxes, or clean spaces, and be officially certified to carry out specific types of biological, chemical, or physical research. For example, research with lasers, explosive chemicals, or pathogenic microorganisms requires specialized laboratories.

## Safety Issues

In some areas of science, you are expected to work with chemical reagents, biological substances, or experimental procedures that are potentially dangerous. For instance, the reagents may be poisonous, radioactive, flammable, or explosive. Equipment may be operated at high pressures, temperatures, or speeds, or it may involve lasers, which pose a threat to the health of you and others when used inappropriately. It is therefore critical to establish any potential safety issues with your experiments before carrying them out (Fig. 4.1). You should then identify the proper way to handle any reagents, biological materials, and equipment, such as wearing appropriate gloves, goggles, breathing equipment, lab coats, shoes, and clothes. You should also have a detailed plan about what you would do if something went wrong (such as a chemical spill, fire, or getting toxic chemicals or pathogens on your skin or in your eyes) before it goes wrong. All researchers should take health and safety training before working in a laboratory and should take refresher courses on a regular basis (usually annually). You may also want to have the telephone numbers of any health and safety, fire, or police officials at hand so you can quickly reach them in the case of an emergency.

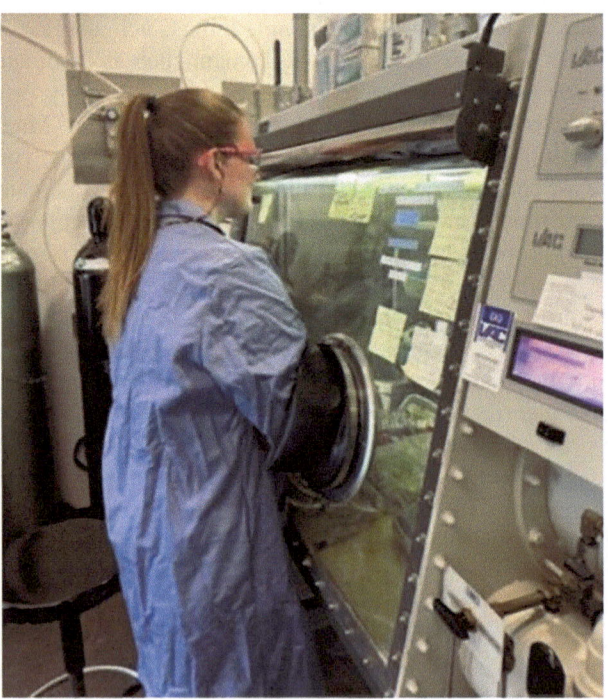

**Fig. 4.1** It is important to wear appropriate personal protection equipment for your experiments. Photograph of one of the authors (IFM) working in the lab

# Good Laboratory Practices

The development of good laboratory practices is critical for the success of your own research, as well as that of the other members in your laboratory. Professors often have people working in their laboratories who do not clean up after themselves, leave workspaces dirty, do not switch off instruments, or do not inform them when supplies are running low or equipment is broken. This adversely impacts everyone working in the laboratory. It is therefore important to develop good laboratory practices to ensure that the research goes smoothly, efficiently, and safely. We have summarized some of the most important practices here:

- *Tidiness and cleanliness*: You should always keep your laboratory space clean and organized, as this makes it easier to locate the materials and equipment you and others need to carry out the research, as well as helping to avoid accidents. If you spill anything on a bench top or piece of equipment (like a balance), then you should make sure to clean it up as soon as possible. You should not expect other people to clean up after you – this is disrespectful, wastes their time, and your spill could contaminate their experiments.
- *Safety*: As mentioned earlier, you should always be aware of all potential safety issues and should wear appropriate personal protective equipment (PPE), such as lab coats, gloves, safety glasses, and closed-toe shoes. You should also dispose of any hazardous waste generated in your research according to the appropriate regulations in your institution.
- *Standard operating procedures*: It is usually important to develop standard operating procedures (SOPs) that provide step-by-step instructions for performing experiments consistently and safely. These SOPs should be followed carefully when you are conducting your experiments. The SOPs can be uploaded to a shared file to allow new lab members to learn techniques more easily.
- *Labeling*: It is important to properly label any chemicals and samples that you are using in your research because this lets you and other members of your lab know what is being used. This helps to avoid mix-ups (such as someone using the wrong reagent or sample) and is important from a safety perspective (if something gets spilled, it is important to know what it is). A label should typically contain the full name of the chemical or sample, the person responsible for it, and the date.
- *Equipment maintenance*: It is important to have a plan to maintain and calibrate the equipment in your laboratory to ensure that it is working properly and produces consistent and reliable results. It is often useful to keep a logbook with each key piece of equipment that people can use to sign up to use it, as well as to report any issues they find. Every time someone uses the equipment, they should sign the logbook. Then, if a problem does occur with the equipment, it may be easier to identify when it occurred and what caused it.
- *Supply stocks*: Typically, you will need a whole range of general laboratory supplies to carry out your experiments, including pipettes, beakers, and paper towels. It is important to ensure that these are always available, or it will negatively

impact your ability to carry out your work. If the supplies are running low, be sure to order some more in good time, or let the person responsible for ordering know.

- *Training*: It is important that you are familiar with all of the key equipment used in your laboratory, as well as the appropriate safety procedures. For this reason, you may have to participate in training programs to learn these procedures and to remain current. If in doubt when using equipment, it is always better to ask someone for guidance, as incorrect usage can lead to expensive and lengthy repair processes. In a lab with many people, it is often good practice to make one or two individuals responsible for each major piece of equipment. These people can help ensure the equipment is properly maintained and operated, as well as providing training to new users.

- *Communication:* It is important to clearly communicate with others in your lab to ensure research runs smoothly and any issues that arise are quickly identified and addressed. This typically involves connecting with the other people in your laboratory, as well as the laboratory manager and professor. Online shared documents are often useful for doing this (such as Google Docs, Slack, or Teams), as these can be used to report chemical lists, consumables needed, and instrument status.

- *Record keeping*: It is important to keep detailed and accurate records of your research, which usually involves using a digital or physical laboratory notebook. You should record the experiments you are doing, the reagents you are using, any problems you encountered, and the results you obtain (see later for more details).

Following good laboratory practices is essential to the running of a safe and productive research program. These skills will be important throughout your career, whether you work in academia, industry, or government.

## Preliminary Experiments

Some researchers are so keen to start their projects that they dive straight into the experiments and start generating lots of data. However, if the experiments are not designed properly or the reagents and equipment used are inappropriate, then all the data obtained may be meaningless. This can be frustrating and disheartening, as well as a waste of time and resources. Conversely, other researchers find it difficult to start their experiments because they want to figure out every detail before they begin or because they are intimidated by the amount of work involved, which can lead to excessive delays. There is therefore a balance between diving in too fast and pondering too long. As discussed in the last chapter, having a systematic strategy (the "manuscript-focused" approach) for planning the research can help with optimizing the experimental design process.

One of the most important elements of the research process is to carry out some preliminary experiments (Fig. 4.2), as this can help to avoid unforeseen errors that might occur from diving into the research too early, as well as to avoid

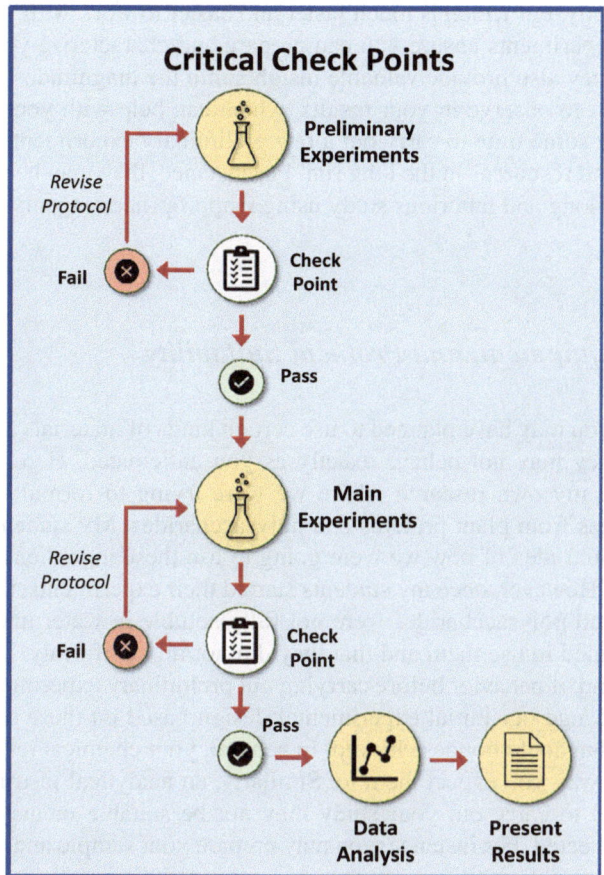

**Fig. 4.2** An experimental study typically requires some preliminary experiments to be performed to establish the appropriate materials and methods. During the main experiments, it is also important to have some critical check points to ensure that the research is proceeding as planned

procrastination by starting with a few simple experiments first. The knowledge gained from these preliminary experiments can save considerable time, energy, and resources. Preliminary experiments are important for several reasons, which are briefly highlighted in the following sections.

## *Optimizing Study Design: The Critical Importance of Preliminary Experiments*

Preliminary experiments help to establish the most suitable system to work on, as well as to iron out any potential problems before embarking on the main study. Typically, they involve focusing on a simple model system that represents the core of the main study, but which is much faster and easier to work with. Well-designed preliminary experiments ensure you can prepare and characterize your samples as anticipated. They also provide valuable insights into the magnitude of any changes you can expect to observe in your results, which can help with your experimental design. Taking some time to carry out a few preliminary experiments can save you a lot of time and resources in the long run. For instance, they may help you to avoid carrying out a long and laborious study using inappropriate reagents or experimental methods.

## *Reagent, Equipment, and Protocol Suitability*

Even though you may have planned to use certain kinds of materials and methods in your study, they may not behave exactly as you anticipated. Here, I will use an example from my own research where we were trying to formulate plant-based meat substitutes from plant proteins and polysaccharides. My students and I came up with an initial idea of how we were going to use these ingredients to create the meat analogs. However, once my students started their experiments, they found that the proteins and polysaccharides were not fully soluble in water under the conditions we intended to use them and that they did not mix uniformly. It is difficult to predict this kind of behavior before carrying out preliminary experiments. We therefore had to change our initial experimental design based on these results because our original plan could not be achieved. In general, your chemical reagents may not behave in the way you expect them to. Similarly, an analytical instrument that has been proposed to carry out your study may not be suitable or may not give the results you expected. For instance, you may prepare your sample and determine that the instrument is not sensitive enough to characterize its properties or that there are some unexpected interferences with the analysis. As an example, if you were going to use chromatography to quantify the concentration of a specific substance within a complex mixture, it is possible that the column you intended to use was not powerful enough to separate the various components present or the detector was not sensitive to measure them. Only by carrying out preliminary experiments can you optimize your reagents and experimental procedures. Once you have completed these experiments, you should be in a position to specify precisely which reagents you require for your study, and so you can go ahead and order them if needed.

## *Magnitude of Changes*

Preliminary experiments can also help to determine the possible magnitude of the changes you might expect to observe in your study. Ideally, the magnitude of the changes should be large enough to be measurable using the methods available. If you have formulated a specific hypothesis, then you want to carry out experiments that lead to either its acceptance or rejection, which means your experiments should be designed so that statistically significant changes can in principle be observed.

Preliminary experiments help to establish whether the expected change in some property is greater than the natural variation expected in this property between samples. If the natural variation is much greater than the expected change, then you will be unable to obtain results that are statistically significant. Obviously, you don't want to spend months carrying out an experiment only to find out your results are meaningless. The variation in results may be inherent in the system being studied. For instance, if you were measuring the influence of eating a particular food (such as spinach) on the gut microbiome (*i.e.,* the population of bacteria living in the large intestine) of people, then it may be difficult to see a pronounced effect because everybody's initial gut microbiome is so different. Some of the statistical methods discussed in the next chapter can be useful when designing experiments based on the expected variability in the parameter being measured. For instance, these methods can be used to establish how many repetitions of an experiment you need to get statistically significant results.

Any observed variations in the results obtained in your preliminary experiments may also be due to an inappropriate experimental design, so that they can be reduced once they have been identified. As an example, in my own research, we were trying to create meat analogs from plant-derived ingredients. We mixed aqueous solutions of plant proteins and polysaccharides together and then heated them to form gels. We then measured the mechanical strength of these gels using a texture analyzer. However, we found that there were large standard deviations in the results – every time we made the measurements, there were large differences in gel strength from sample to sample. Consequently, it was difficult to see any clear trends in the data when we varied some experimental parameter (such as pH or salt concentration). We eventually found that this problem was due to the presence of air bubbles that were incorporated into the samples during their preparation. The number and size of these air bubbles varied between samples, which led to large changes in their mechanical properties. This problem could be overcome in several ways: (i) changing the preparation method to avoid air bubble formation; (ii) removing air bubbles after making the protein-polysaccharide solutions; and (iii) standardizing the preparation procedure so that air bubble formation was the same in all samples. By identifying this problem during the preliminary experiments, we could then design the main experiments to reduce any potential sources of deviation.

After carrying out your preliminary experiments, you should have become familiar with the different reagents, equipment, and protocols you intend to use in your main study, as well as any potential problems that could arise. You are then in a much better position to plan your main study so as to get reliable and consistent results.

## Main Experiments

Once you have decided on a good system to work on and have ironed out as many potential problems as possible through your preliminary experiments, it is time to start your main experiments (Fig. 4.2). At this stage, it is important to be highly rigorous, well-organized, and systematic when performing your experiments and recording your data. In this section, we provide general suggestions that should help you to carry out your experiments efficiently and effectively.

### *Develop a Clear Experimental Plan*

You should create a clear and concise plan for carrying out your experiments. This will be based on the design of your scientific manuscript described in the previous chapter. The experimental plan might look like a recipe used by cooks to create a meal. It should contain detailed instructions on how the samples will be prepared and characterized. It may also contain a checklist of all the different experiments that must be carried out to complete the research. The plan should include several *standard operating procedures* (SOPs), with each one giving a detailed description of how a particular set of experiments is performed (Fig. 4.3). For instance, in my research group (DJM), we have an SOP for measuring the viscosity of foods using a rheometer, another one for measuring the microstructure of foods using a fluorescence microscope, as well as various others. The overall experimental plan could become the "Materials and Methods" section in the scientific manuscript or thesis

---

### SOP: Thermogravimetric Analysis (TGA)

1. Turn on computer
2. Open gas valve, turn on the machine (always open gas valve first), open the software
3. Press arrow button to open the machine
4. Calibrate cell weight (Carefully place cell in instrument, it breaks easily and takes weeks to fix!)
5. Take the cell and put your sample in it (don't overload, it should be lower than 2/3rds total volume)
6. Sample loading shape could affect the result so push the sample gently into the cell
7. Select or define operating procedure (start temperature, end temperature, heating rate)
8. Start run
9. Export data when finished.
10. Wait for temperature of the instrument to drop to room temperature.
11. Remove sample.
12. Turn off the machine and close gas valve

**\* Cleaning cells (should be done in hood)**
Cells are reusable so you should be cleaned
- Take lighter and heat cell from bottom while holding it with a clamp.
- Heating will burn off any residual sample leading to a clean cell
- Place all items back in their original place

**Fig. 4.3** Example of a standard operating procedure (SOP) for how to carry out thermogravimetric measurements. Typically, there will be several SOPs needed for your proposed research project

chapter based on your experiments. Wherever possible, everything should be standardized in your experiments such as the order of addition, mixing, reaction, and storage conditions (*e.g.,* pH, ionic strength, mixing speeds, times, temperatures, oxygen levels, light levels *etc.*), and instrument operating parameters (*e.g.,* model, settings). Once you have established the experimental plan and associated SOPs, and it has proved robust in your preliminary experiments, then you should rigorously stick to it. If you change some of the parameters every time you do your experiments (such as the pH, mixing speed, incubation temperature, or storage time), then your results may vary, and you may not know the reason why.

## Data Collection and Storage

As you carry out your experiments, you generate data that must be carefully collected, collated, and stored. It is important to have a clear plan about how you intend to do this. In this section, we highlight some of the important aspects of data collection, collation, and storage that you should consider.

### *The Importance of Keeping a Good Notebook*

Many professors require their students to keep detailed experimental notebooks, especially those who are working in areas of science that are highly competitive and may lead to publications in high-quality journals or to patent applications. In addition to recording the data, the notebook indicates when the experiment was performed, who did it, and what procedures were used, which can be important when claiming a new discovery.

For each study, a laboratory notebook should contain several pieces of information:

- Your name.
- The time and date the experiments were performed.
- The nature of the chemical reagents, protocols, and equipment used.
- The detailed results of the study, including the raw data collected for each repetition.

You may record this information in a physical or digital notebook. When performing your experiments, you might think you will always remember exactly what you did, but it is very easy to forget. Graduate students are busy with many tasks and may work on the same project for several years. If you do not include all of the above information, you may not remember precisely how your experiments were performed, which can make it difficult to repeat or interpret them. It may seem time consuming to include this kind of information every time you start a new study, but it is an important thing to do. Your notebooks should be detailed and well organized so that other people can easily read and understand them. You will certainly thank yourself for doing this when you come back to look at your experimental results after a few months or years.

## *Quality Control Checks*

When carrying out the main experiments, many students adopt a relatively passive attitude to performing their research and collecting their data, without actively assessing whether the study is going as expected. We find that it is useful to analyze and plot the data from an experiment as it is proceeding. Then, if an unexpected problem arises (such as a wrong chemical reagent being used or an analytical instrument malfunctioning), it may be possible to quickly identify it and eliminate it. Nevertheless, these quality control checks should not be used to eliminate data that do not conform to your expectations. It is always important to keep an open mind because unanticipated results may be an important new discovery (see later).

## *Data Backup*

It is always important to back up your data. Most data are now in a digital format, so it is important to back it up to the cloud, as well as download a copy onto an appropriate memory drive. During my career as a professor (DJM), I have had several occasions when students have come to my office nearly in tears because their computer broke or got a virus, and they lost all their data. You do not want to spend weeks, months, or even years collecting high-quality data and then see it all vanish because you did not back it up properly. Many universities and private companies (such as Apple or Google) have cloud-based storage facilities where you can safely store your data. You should develop a routine of backing up your data periodically (such as every day, week, or month). Ideally, you should store all the data from a particular study in one place when it is completed. Then, you can easily access all of the information you need to prepare a scientific manuscript, presentation, or thesis. Some government agencies require the recipients of their grants to make any data generated by a funded project publicly available so that anyone can utilize it. In this case, you should store your data in a place and format that other people can easily access and understand.

## Quality Control of Experiments: Critical Check Points

During my career as a professor (DJM), there have been several instances where a graduate student who has been working on their experiments for months comes into my office to show me their data. As we are discussing their results, it becomes clear that there was some mistake in the initial experimental design that made their results inaccurate or meaningless, so they could not be used. The student then had to perform all the experiments again. This was extremely discouraging for the student and wasted lots of time, money, and resources. It is therefore important to avoid this situation whenever possible.

There are many reasons why experiments can go wrong or do not go as expected. Sometimes, the science is extremely complicated, and it is challenging to get the experimental design correct the first time. It may require a few iterations of an experiment to perfect the protocol. However, this may also be because the experimental design was not thought through properly before starting the experiments or because the experiment was not set up or performed correctly. A number of common reasons for experiments going wrong are highlighted in Table 4.1.

From our experience, taking time to perform meaningful preliminary experiments and to carefully design and refine the main experimental protocol saves considerable time, energy, and resources in the long run by eliminating unnecessary

**Table 4.1** Some common errors that lead to experiments going wrong

| Source of problem | Description |
| --- | --- |
| Bad design | The initial experimental protocol was not carefully planned. The design was unsuitable for addressing the problem of the study. |
| Preliminary experiments issues | Preliminary experiments are often carried out before the main experiments to establish if the reagents, protocols, and instruments used will provide reliable, consistent, and meaningful results. It is important to have a CHECK POINT at the end of the preliminary studies, where you critically assess the suitability of the experimental protocol proposed for the main study. |
| Reagent issues | It is important that any chemical reagents (including water) are appropriate, have the required purity, have been stored under proper conditions, and have not passed their use-by date. If the chemicals used are inappropriate, the results obtained will be meaningless. |
| Instrument issues | It is important that any instruments used in your experiments are clean and functioning properly (such as pH meters, balances, microscopes, spectrometers, chromatography devices *etc.*). The instruments may need to be cleaned and calibrated before you start. If the instrument is not working properly, all your results will be meaningless. |
| Protocol issues | It is important that any standard operating procedures (SOPs) are suitable for the systems you are working with, and that they have been optimized for your experiments. For instance, times, temperatures, orders of addition, and stirring speeds, may all have to be specified in a SOP and rigorously followed for all samples. You may have to perform them several times before starting your main experiments to ensure they are appropriate. This can be done in the preliminary experimental stage. |
| Interpretation issues | It is important to carefully record, collate, and interpret the results of your experiments. Errors can creep in during the recording and interpretation of data. It is therefore important to keep detailed records of all your experiments and calculations. When using spreadsheet programs (like Excel), it is possible to make mistakes when typing in equations or formulas. It is therefore important to carefully check your data entries and calculations. |

**Table 4.2** Example of a quality control plan with critical check points that can be used to reduce errors and improve the efficiency of your research

| Source of problem | Description | Check |
|---|---|---|
| **Check point 1** | | |
| Reagents | Do you have all the reagents required? | ☐ |
| | Are they of the required purity? | ☐ |
| | Are they within their use-by date? | ☐ |
| | Are there any potential hazards? | ☐ |
| | Are there any special usage conditions? | ☐ |
| Instruments | Do you have all the instruments required? | ☐ |
| | Are they available? | ☐ |
| | Are they clean and functioning properly? | ☐ |
| | Have you been trained to use them? | ☐ |
| | Do they need to be calibrated? | ☐ |
| Protocols | Are your experimental protocols appropriate for your study? | ☐ |
| | Have you been trained to perform them? | ☐ |
| | Is there an SOP you need to use? | ☐ |
| **Check point 2** | | |
| Preliminary experiments | Do your preliminary experiments provide results that indicate that the reagents, instruments, and protocols used will provide reliable and consistent results? | ☐ |
| **Check point 3** | | |
| Experimental Design | Have you carefully designed your study so that if the experiments are successfully performed, they will provide meaningful data that will tell a good story (leading to a scientific manuscript)? | ☐ |
| **Check point 4** | | |
| Interpretation | Has all the data been tabulated and analyzed correctly? | ☐ |
| | Have you double checked that your calculations are correct? | ☐ |

errors. You should not blindly follow advice given by other lab members. Instead, you should make sure you fully understand any experimental protocols you are using. Problems can arise from lab members passing along information that becomes increasingly inaccurate over time, so people end up following bad advice.

One way of avoiding unnecessary errors in your research is to develop a good *quality control* plan that has several *critical check points* where you stop and assess the validity and quality of your results (Fig. 4.2 and Table 4.2).

## The Importance of Being Open Minded

Another important characteristic of a good scientist is to be receptive to the importance of any unanticipated results. If an experiment is designed that consistently produces an unexpected result, it is important to be aware that the result may be due to poor experimental design or that it may be due to some interesting new phenomenon. Many of the most interesting discoveries in science are made by researchers who observe a phenomenon that did not correspond to their original expectations, and then follow up on it. In these situations, it is important to perform further experiments to provide additional evidence about the factors influencing the observed phenomenon, as well as to establish its origin.

The discovery of penicillin is a classic example of a scientific discovery that was made due to an unexpected observation, that was then further investigated. In 1928, Scottish scientist Alexander Fleming was conducting experiments with bacteria when he noticed that a mold called *Penicillium notatum* had contaminated one of his petri dishes. He observed that the bacteria in the area surrounding the mold were not growing and realized that the mold was producing a substance that inhibited bacterial growth. This unexpected observation led Fleming to further investigate the mold and isolate the substance responsible for its antibacterial properties, leading to the discovery of penicillin. This unexpected finding resulted in the discovery of a potent antibiotic for treating a broad range of bacterial infections, thereby saving countless lives.

## Concluding Remarks

Scientific studies are often time consuming, laborious, and expensive to carry out. It is therefore important to perform them as efficiently and reliably as possible. Ideally, you do not want to make unnecessary mistakes, as they will waste your time, money, and resources. Therefore, it is important to carefully plan your experiments before starting them. Spending a few hours planning can save weeks or months performing research that is inaccurate or irrelevant. It is also important to establish quality control checks throughout your studies, as they can help you rapidly identify any problems that arise. You can then rectify these problems quickly before you waste too much time.

**Key Points**
- Research typically involves several steps: problem identification and definition, experimental design, data collection, data analysis and interpretation, and communication.
- Before starting your experiments, ensure you have all the chemicals, reagents, and equipment available and you know how to use them properly.
- Understand the risks of each chemical, reagent, and piece of equipment used in your studies. Use this knowledge to establish an emergency plan.
- Follow good lab practices, including tidiness, safety, labeling, equipment maintenance, standard operating procedures, supply stocks, training, communication, and record keeping.

(continued)

(continued)

- Preliminary experiments are important for setting the framework of your research. They help to identify appropriate chemicals, reagents, and equipment, as well as optimize protocols, identify problems, and indicate the kinds of results that can be expected.
- Prepare a detailed experimental plan and standard operating procedures for your study, which should be clear, precise, and accurate, so you and others may repeat your work. It is also important to record this information clearly into a lab notebook.
- Perform quality control checks during your research to reduce the chance of errors. This will save you time, as you can catch mistakes early rather than having to repeat months' worth of experiments.

## Resources

Creswell JW, Creswell JD (2018) Research design: qualitative, quantitative, and mixed methods approaches, 5th edn. Sage, London
Wilson EB (1991) An introduction to scientific research. Dover Publications, London

# Chapter 5
# Data Analysis and Reporting: Presenting Your Best Face

## Introduction

Once you have completed your experiments, you will have generated a lot of data. Now you have to analyze it, and present your research findings in a clear, informative, and impactful manner. The world is a complex and dynamic place, so there are natural variations in the physical, chemical, and biological phenomena studied by scientists. As a result, there will be inherent variations in your data. Consequently, you must carry out enough measurements or observations to obtain reliable

D. J. McClements et al., *How to be a Successful Scientist*,
https://doi.org/10.1007/978-3-031-51402-9_5

estimates of the value being measured to account for this natural variation. In this section, we provide a brief overview of some of the most common statistical approaches used for analyzing the data obtained from scientific studies. For more detailed information, it is useful to refer to dedicated books on experimental design and statistics, such as the ones listed in the resources section of this chapter.

Before starting, it is useful to define some of the terminology commonly used in statistics to facilitate the application of this important tool for designing and interpreting experiments:

- *Population*: The term *population* is typically used to refer to the entire group of things (data, objects, people, or events) a researcher is interested in, which typically share some common feature (Fig. 5.1). For instance, if a researcher is studying obesity in the USA, then the population of interest would be all the people who live in that country.
- *Sample*: The term *sample* is typically used to refer to a subset of the overall population (Fig. 5.1). In some cases, this may be a *random sample* that is assumed to have properties that accurately reflect those of the entire population, provided the fraction collected is sufficiently large and collected carefully (Fig. 5.1 **left**). For instance, you may want to estimate the percentage of people who are obese in the USA, but you cannot measure the height and weight of everyone in the country. For this reason, you only measure these quantities for a much smaller fraction of the population (perhaps a few thousand people) and then extrapolate from that information to the entire population. In other cases, the subset you select may be a *nonrandom sample* whose properties may or may not reflect those of the entire population (Fig. 5.1 **right**). For instance, the subset of samples that are of interest may be those people who are vegetarians. In this case, you may want to establish whether vegetarians are more, less, or as obese as the entire population, which can be achieved using appropriate statistical tools.

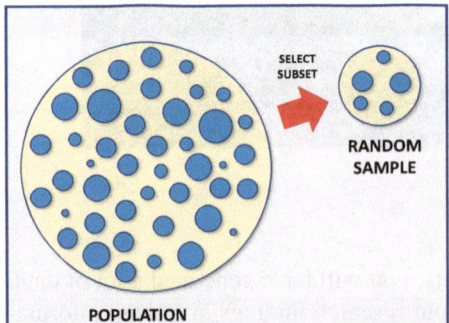 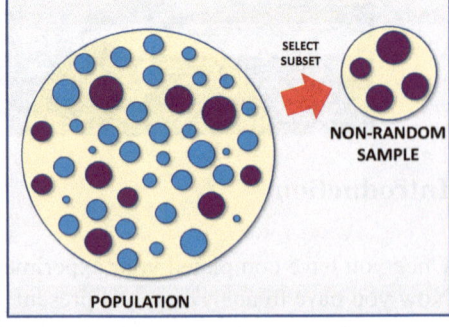

**Fig. 5.1** In some studies, a researcher may randomly select a sample from a population (left). Ideally, the properties of this random sample should accurately represent those of the whole population. In other studies, a researcher may select a nonrandom sample from a population (right) and want to establish whether its properties are similar or different from that of the population. In the example shown here, the average size of the purple circles is larger than those in the overall population

It is important to clarify the reason that statistics are being used when analyzing the results of your studies, as this influences the kind of statistical methods you need to employ. In general, there are two main types of statistical approaches used by research scientists:

- *Descriptive statistics*: In this approach, a researcher simply uses statistics to analyze a particular dataset, which has usually been collected from experiments or observations, to obtain some reliable estimate about their properties. For instance, statistics may be used to represent the entire dataset in terms of a few simple parameters, such as the average, the extent of variability, or the fraction above or below some critical value. As an example, this approach might be used if you wanted to find the average height of the members of a basketball team, as well as the fraction who were over 6 foot 7 inches (200 cm) tall.
- *Inferential statistics*: In this approach, a subset of data is collected from a larger population (Fig. 5.1), and the aim is to infer information about the entire population from this subset. For instance, you might want to determine the average and variability of a quantity of interest for the entire population. You can never know the true values for the entire population because you are not analyzing every sample. However, statistics can be used to characterize the properties of the sample you have collected and to provide some insights into how close these values are expected to represent those of the entire population. Selecting a representative subset is particularly important in inferential statistics. For example, the United States Centers for Disease Control and Prevention (CDC) estimates the national obesity rate from measurements taken on a small fraction of the entire population because it is impractical to measure the weight and height of everyone in the country. However, it is important that the sample size is large enough and that the subset selected for analysis accurately represents the entire population; otherwise, the results will be unreliable and meaningless.

An appreciation of the difference between descriptive and inferential statistics is important for research scientists. Researchers often assume they are using simple descriptive statistics when analyzing the data collected from their experiments or observations, when in reality, they are using inferential statistics. For instance, a scientist may be trying to measure the loss of vitamin D in milk when it is pasteurized. The scientist prepares a number of samples and measures the vitamin concentration before and after pasteurization and then calculates the percentage that has degraded. They may perform the measurement several times and obtain slightly different results (*e.g.*, 24.3, 29.4, 19.8, and 22.4% loss) due to natural variations in the samples collected and the methods used to analyze them. If they could make an infinite number of measurements on the samples, there would be a distribution of data around some central value, with the spread of the data depending on the natural variability of the system being studied. This infinite set of measurements can be considered to be the entire population of measurements and could be taken to represent the true properties of the thing being analyzed. In many cases, the variations in the data are random, and the entire population can be represented as a normal distribution (Fig. 5.2). A scientist therefore wants to design their experiments so that

**Fig. 5.2** Plots of the likelihood (frequency) that a certain value will be measured assuming a normal distribution. These two populations have the same population means ($\mu$) but different population standard deviations ($\sigma$)

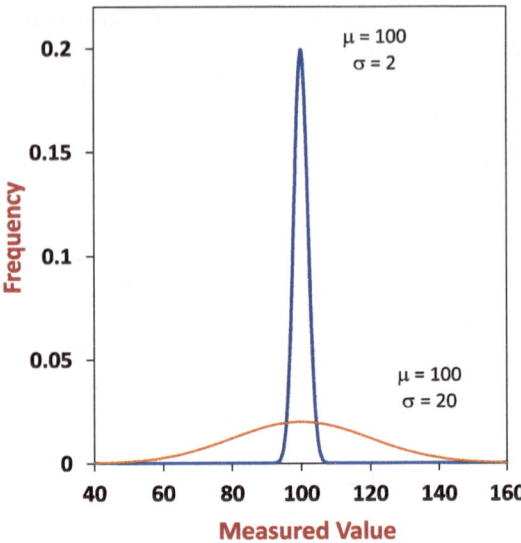

the number of measurements carried out provides results that accurately represent the true properties of the entire population. They also want to know how confident they are that the values calculated from their limited number of samples agrees with the true values.

In the remainder of this section, we present several statistical tools commonly used to facilitate the design and interpretation of scientific studies. The adoption of these tools can help you obtain results that are meaningful and reliable.

## Measures of Central Tendency

It is often convenient to represent a dataset by a single parameter, which provides insights into the central or most frequent value. In statistics, the three most common measures of central tendency are the mean, median, and mode, which each have their own advantages and disadvantages for representing the properties of a population.

## *Mean*

The *mean* is widely used to provide an indication of the average properties of a dataset. A sample mean can be calculated using the following expression:

$$\bar{x} = \frac{x_1 + x_2 + x_3 \ldots x_n}{n} = \frac{\sum\limits_{i=1}^{n} x_i}{n} \tag{5.1a}$$

Here, $n$ is the number of objects in the sample, $x_i$ is the value measured for each object, and $\bar{x}$ is the calculated sample mean. The sample mean provides the *best experimental estimate* of the value that can be obtained from the limited number of measurements that have been made. It does not necessarily correspond to the *true* mean that would be obtained if you could make an infinite (or very large) number of measurements.

The true mean of the overall population can be calculated using a similar equation, but it is given a different symbol ($\mu$) to distinguish it from the sample mean ($\bar{x}$):

$$\mu = \frac{\sum_{N}^{i=1} x_i}{N} \tag{5.1b}$$

Here, $N$ is the total number of objects in the entire population, $x_i$ is the value measured for each individual object, and $\mu$ is the population mean.

There are several reasons why the sample mean may be different from the true population mean. First, you may not have made enough measurements to obtain a reliable estimate. For example, imagine you wanted to determine the probability of getting "heads" when tossing a coin, you might toss the coin numerous times and record the percentage of them that comes up heads. If you only tossed the coin four times, you may get one head and three tails, so you calculate a sample mean of 25% for getting "heads." However, if you tossed the same coin hundreds or thousands of times then you are likely to get a value much closer to the expected value of 50%. Consequently, it is important to perform a sufficiently large number of measurements so that the calculated sample mean provides a reasonably accurate representation of the true mean. On the other hand, you do not want to make more measurements than needed to get a good estimate because this is time consuming, laborious, and expensive. An estimate of how closely your measured sample mean agrees with the true population mean can be obtained by calculating the *standard error* and/or *confidence interval*, which is discussed later. The measured sample mean may also not agree with the true mean because of systematic errors in the way you collected or characterized your samples. Consequently, it is important to identify and eliminate these errors (see later).

How close the measured mean agrees with the true mean represents the *accuracy* of the measurements. The accuracy can be quantified in terms of the absolute error ($E_{abs}$), which is the difference between the measured mean ($\bar{x}$) and the true mean ($\mu$): $E_{abs} = (\bar{x} - \mu)$. Ideally, this parameter should be as small as possible. In practice, it is difficult to establish the accuracy of your measurements because the true mean is usually unknown.

## Median and Mode

Other commonly used measures of central tendency are the median and mode. The median is the middle value in a dataset when it is arranged in order from the lowest to highest. In other words, 50% of the data have a value below the median, and 50%

have a value above the median. The median is useful in situations where a few outliers disproportionally affect the calculated mean. For instance, if you wanted to get an idea of the typical wealth of the people living in a small town, then calculating the mean may be inaccurate because the wealth of a few multibillionaires who lived there contributed disproportionately. In this case, the median may be more useful. As an example of how to determine the median from a set of measurements, consider the following dataset: 7, 2, 9, 4, 1, and 5. First, the measurements should be arranged in sequence in ascending order: 1, 2, 4, 5, 7, and 9. Since there are six measurements in total, the median is the middle value or values, which in this case are the numbers 4 and 5. The median is then calculated from the average of these two numbers = 4.5.

The mode is the value that appears most frequently in a dataset. For instance, the mode of the dataset 1, 1, 2, 3, 3, 3, 3, 4, and 5 is "3" because it appears four times. In some situations, a dataset can have multiple modes because two or more values appear with the same frequency, as in the dataset 1, 1, 2, 2, 2, 3, 3, 3, 4, and 5, where "2" and "3" both appear three times. The mode is useful when one wants to establish the most frequently occurring value or values in a dataset. For instance, a researcher may conduct a survey to determine which is the most popular color for a candy.

## Measures of Spread of Data

As mentioned earlier, when you are making repeated measurements or observations of some phenomenon, your results typically differ to some extent due to the natural variability of the world. For instance, you may prepare a polymer-based material in the laboratory four times using the same chemical reaction and then measure the hardness of each sample, which leads to results of 4.34, 4.12, 4.84, and 3.98 N. This natural spread in the data may occur for several reasons, such as variations in the initial chemicals used (such as monomers or solvents), sample preparation conditions (such as time, temperature, or stirring speeds), laboratory conditions (such as temperature and humidity), or sample characterization conditions (such as changes in the instrumental operating conditions used to measure hardness). The *spread of the data* provides a measure of how close together repeated measurements are to each other. It therefore provides insights into the *precision* of the data. Typically, the lower the spread is, the easier it is to obtain a reliable estimate of the true mean value. For example, if the hardness values for your polymer materials were relatively close together (*e.g.,* 4.30, 4.28, 4.29, and 4.31 N), then you are more likely to calculate an accurate mean value from a limited number of experiments than if they are far apart (*e.g.,* 3.50, 4.88, 2.99, and 5.21 N). However, the accuracy of your calculated mean can be reduced by carrying out more measurements. Statisticians have developed several ways of quantifying the spread of the data obtained from experimental studies, as well as statistical tools to predict how close the measured mean is to the true mean based on the known spread of the data.

## Standard Deviation

The *standard deviation* is the most common measure of the spread of experimental measurements or observations. It is determined by assuming that the values vary randomly around the mean in the form of a normal distribution (Fig. 5.2). The standard deviation (SD) of a sample can be calculated from a limited set of experimental measurements using the following equation:

$$SD = \sqrt{\frac{\sum\limits_{n}^{i=1}\left(x_i - \bar{x}\right)^2}{n-1}} \tag{5.2a}$$

The standard deviation of the entire population can be calculated using a similar equation, except that a different symbol ($\sigma$) is used to express this parameter:

$$\sigma = \sqrt{\frac{\sum\limits_{N}^{i=1}\left(x_i - \mu\right)^2}{N}} \tag{5.2b}$$

Here, $n-1$ appears in the denominator of the equation used to calculate the sample standard deviation (rather than $n$) because only a limited number of samples are selected from the entire population. As a result, the value of the sample standard deviation has a greater error (and therefore range) than the population standard deviation.

The larger the standard deviation is, the broader the variability in the data (Fig. 5.2), which has important consequences when designing experiments and establishing significant differences between samples (see later). It should be noted that the standard deviation has the same units as the mean, so they can be reported together. Another statistical expression that is closely related to the standard deviation is the *variance*, which is the standard deviation squared.

## Standard Error of the Mean

The *standard error of the mean* (SEM) provides an estimate of how close the sample mean determined from a limited subset of data is to the true mean. It is typically calculated as the standard deviation of the population divided by the square root of the sample size:

$$SEM = \frac{\sigma}{\sqrt{n}} \tag{5.3}$$

A smaller SEM indicates that the sample mean is more likely to be closer to the population mean. In reality, one may not actually know the standard deviation of the entire population, so an approximate SEM value is calculated by using the standard deviation calculated from the selected sample: $SEM = SD/\sqrt{n}$ . It is important not to confuse the standard deviation and the standard error. The SEM provides an estimate of how far the sample (measured) mean is likely to be from the population (true) mean. In contrast, the sample standard deviation provides an estimate of the spread of the data points around the sample mean.

## *Confidence Intervals*

The confidence interval (CI) is related to the standard error and can be useful when designing and interpreting experiments:

$$CI = \frac{z\sigma}{\sqrt{n}} = z \times SEM \tag{5.4}$$

Here, $z$ is the confidence interval level, $\sigma$ is the standard deviation of the population, and $n$ is the number of measurements. The confidence interval tells you the level of confidence you can have that the sample mean measured from your subset of $n$ test samples falls within $\bar{x} \pm CI$ of the true mean of the entire population. To have a 95% confidence level, the $z$ value should be equal to 1.96. In other words, there is a 95% chance that the measured mean would fall between $\bar{x} \pm$ 1.96 × SEM ≈ 2 × SEM. Ideally, the lower the confidence interval is, the closer the sample mean calculated from a limited number of measurements is to the true mean. This equation shows that the confidence in your mean value improves as the number of measurements used to calculate the sample mean increases. The more measurements you make, the better your estimate.

The impact of the standard deviation and number of measurements on the confidence interval is shown in Fig. 5.3. It is assumed that the true mean has a value of 100. In one set of experiments, the standard deviation is taken to be quite small ($\sigma = 2$), but in the other set of experiments, it is taken to be quite large ($\sigma = 20$). For a sample set containing $n$ data points, there is a 95% chance that the sample mean will fall between the two similar lines drawn in the diagram (shaded areas). The distance between these lines represents the confidence interval. For instance, if you made five measurements ($n = 5$), then there is a 95% probability that the calculated mean will fall between 98.2 and 101.8 for the narrow dataset ($\sigma = 2$) and between 82.1 and 117.9 for the broad dataset ($\sigma = 20$). A couple of useful things can be learned from the calculations shown in Fig. 5.3. First, if the standard deviation of your measurements is relatively small, then you only need to carry out a few measurements to get a reasonably good estimate of the true mean. Second, your estimate of the true mean improves as the number of measurements you make increases.

**Fig. 5.3** When you calculate the mean value from a limited number of samples, it will fall within a range around the true mean (μ). The smaller the population standard deviation (σ) or the larger the number of measurements made (*n*), the more likely the measured mean ( $\bar{x}$ ) is close to the true mean (μ)

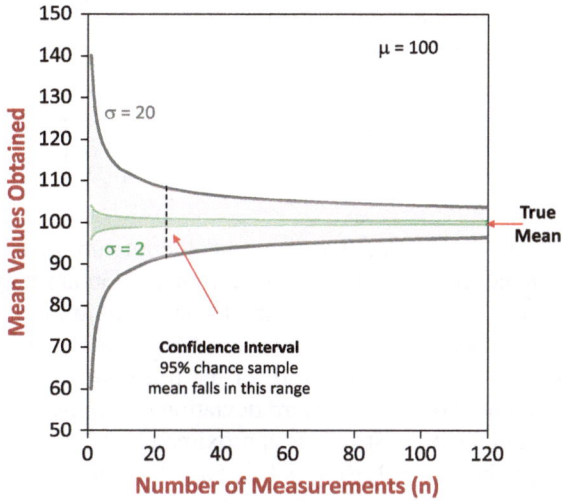

## Coefficient of Variation

The *coefficient of variation* (CV) is used to provide an indication of the *relative* spread of the data around the mean value:

$$CV = [\sigma / x] \times 100\% \qquad (5.5)$$

In this case, the standard deviation is normalized relative to the mean. This is important when comparing the variability of measurements that have different magnitudes. For instance, 0.2 ± 0.1 has a much lower absolute standard deviation than 100 ± 1, but there is much more relative variation in the former number (CV = 50%) than in the latter one (CV = 1%). Thus, the coefficient of variation is useful when comparing the reliability of data with substantially different mean values.

## How Many Measurements Should You Make?

An important factor to consider when you are designing an experiment is to establish how many measurements you should make to obtain a reasonably good estimate of the mean and standard deviation. If you make too few measurements, then they may not provide an accurate representation of the true values. However, you do not want to make too many measurements because that would be time consuming and expensive. So how do you decide? The statistical parameters just discussed can help with this.

For instance, the following equation can be used to estimate the number of measurements needed to obtain a certain level of precision, which is obtained by rearranging Eq. 5.4:

$$n = \left( \frac{z\sigma}{CI} \right)^2 \tag{5.6}$$

In this case, the CI represents the maximum acceptable difference between the sample mean and the population mean you are targeting. As an example, suppose you are trying to estimate the mean concentration of vitamin D in people's bloodstream, and you want to have a 95% confidence level that the sample mean you obtain is within 0.1 ng/mL of the true (population) mean. From previous studies, you know that the standard deviation of the measurements of vitamin D concentration in the bloodstream is approximately 0.2 ng/mL. You can then put the appropriate values ($z = 1.96$, $\sigma = 0.2$, $CI = 0.1$) into Eq. 5.6 and calculate the number of measurements you need to perform to achieve the needed precision: $n \approx 15$. If the known standard deviation of the measurements was smaller, say $\sigma = 0.1$ ng/mL, then the number of measurements needed to achieve the same level of confidence and precision would be substantially less: $n \approx 4$.

The number of measurements you need to obtain an accurate representation of the true mean and standard deviation depends on several factors, including the variability of the data ($\sigma$), the desired level of precision (CI), and the confidence level ($z$) you want to achieve. Typically, the more measurements you make, the better your estimates of the mean and standard deviation will be. Data that naturally have a relatively low variability require fewer measurements than data with a high variability. Consequently, the number of measurements you need depends on the type of experiments or observations you are making, as well as the intended purpose of the study.

In the physical sciences, researchers typically carry out at least three or four repetitions of each experiment. This involves repeating the entire experiment from scratch at least three or four times and collecting the resulting data. This is different from carrying out multiple measurements on a single sample that has only been prepared once. In the latter case, you are mainly determining the variability in the analytical procedure you use to make the measurement, whereas in the first case you are measuring the variability in the whole process, including variations in materials, preparation procedures, and measurement methods. Although many research scientists routinely use three or four repetitions of their experiments, few of them actually try to determine if this is an appropriate number. It may therefore be useful to check this when you are designing your experiments. This requires you to carry out a number of preliminary measurements to obtain an estimate of the standard deviation and then to specify how close you want the measured mean to be to the true mean at a 95% confidence level (Eq. 5.6).

# Establishing Statistical Differences: ANOVA

In many studies, it is important to determine whether the properties of two or more samples are the same or different. For example, you may be carrying out a trial to determine whether taking vitamin supplements changes the vitamin levels in people's bloodstream. You therefore measure the vitamin levels in a group of people who take vitamin supplements for a month and in another group who did not take the vitamin supplements. After calculating the data from the two groups, you found that the mean vitamin concentration in the bloodstream was higher for the people taking the supplements. This could be a real effect of taking the supplements, or it could simply be the result of random fluctuations in the vitamin levels between different people. So how can you tell the difference and be confident that it is a real effect?

There are various statistical methods available for establishing whether experimental measurements are significantly different from each other or not, and it is important to select the most appropriate one for your particular situation. In this section, we focus on just one of these methods to highlight the importance of statistics when analyzing and comparing data.

Analysis of variance (ANOVA) is a statistical method commonly used to establish whether the means of two or more samples are significantly different from each other. Each sample represents a series of repeated data points collected from a group that has something in common. For instance, one sample could be the vitamin concentrations in the bloodstream of all the people who received the vitamin supplement, whereas another sample could be the vitamin concentrations in the bloodstream of all the people who did not take the supplement. In ANOVA, the total variation in the data is divided into two components: (i) the variation between the samples and (ii) the variation within the samples. These two sources of variation are then compared with each other. If the variation between the samples is significantly greater than that within the samples, then one can assume that there is a statistically significant difference between them.

There are different kinds of ANOVA that can be carried out depending on the type of data you have, such as single-factor, two-factor with replication, and two-factor without replication. Here, we give an example that utilizes the simplest one: single-factor ANOVA. These relatively simple statistical analyses can often be carried out using spreadsheet programs (like the *Data Analysis* tool in Excel).

Single-factor ANOVA is used for analysis of variance on data obtained for two or more samples. It tests the hypothesis that each sample is drawn from the same underlying probability distribution against the alternative hypothesis that samples are drawn from different underlying probability distributions. To provide more clarity, some (made up) data for the vitamin study are shown in Table 5.1. Let's assume that the supplement is a multivitamin, and you measure the concentrations of both vitamin D and vitamin E in the bloodstreams of two groups of people, one who take the supplement (test sample) and one that don't (control sample).

**Table 5.1** Representative examples of an experiment to determine whether there was a significant difference between the two groups

|  | Vitamin D experiment (ng/mL) | | Vitamin E experiment (ng/mL) | |
| --- | --- | --- | --- | --- |
| *Measurements* | Control | Test | Control | Test |
| Person 1 | 13.0 | 45.0 | 16.4 | 40.3 |
| Person 2 | 21.4 | 36.3 | 21.4 | 42.1 |
| Person 3 | 35.2 | 19.3 | 22.6 | 39.8 |
| Person 4 | 23.1 | 37.0 | 23.1 | 37.6 |
| **Mean** | **23.2** | **34.4** | **20.9** | **40.0** |
| **SD** | **9.2** | **10.8** | **3.1** | **1.9** |
| p value | 0.16 (not sig. different) | | 0.00 (sig. different) | |

In this example, it is assumed that the vitamin D and E concentrations in blood samples collected from groups of individuals who either did or did not take a vitamin supplement were measured. Four people were assumed to be in each group. A significant difference was defined at a 95% confidence level ($p < 0.05$). The *p* value was calculated using single factor ANOVA (Excel Data Analysis tool)

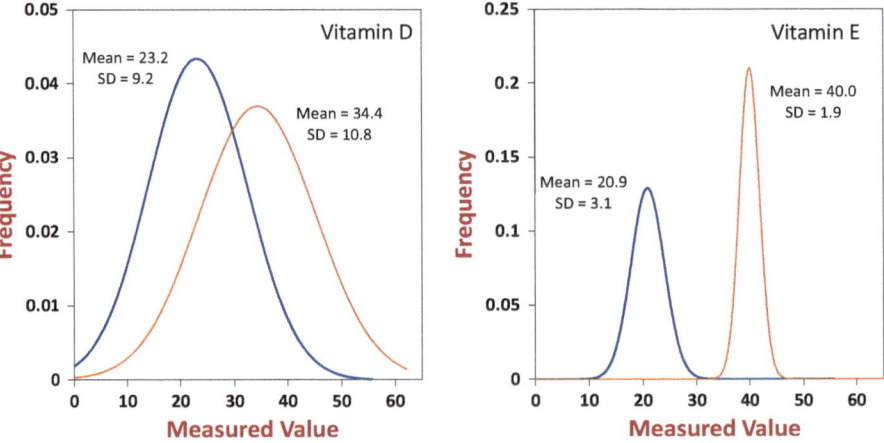

**Fig. 5.4** Estimated distributions of the data from the vitamin study reported in Table 5.1. The thick blue lines represent the people who did not take the supplement, whereas the thin orange lines represent the people who did take the supplement. In this case, it would be much easier to say there was a significant difference between the groups for the vitamin E (little overlap) than for the vitamin D (large overlap)

The mean and standard deviation of each sample were calculated from the data shown in Table 5.1 and then used to plot the frequency distributions assuming that the data followed a normal distribution (Fig. 5.4). It should be remembered that only a limited number of data points are taken for each group, so that the calculated sample mean and standard deviation values are only estimates of the true values. For vitamin D, there was a large overlap between the test and control groups, and it was not possible to say at a 95% confidence level that they were significantly different ($p > 0.05$). In contrast, for vitamin E, there was little overlap between the test and control groups, so it was possible to say that they were significantly different ($p < 0.05$) using the ANOVA test.

We can also see why it is difficult to establish a significant difference for vitamin D, but not with vitamin E by calculating the confidence intervals. For vitamin D, there was a 95% chance that the true mean was between 14.2 and 32.2 ng/mL for the control sample and between 23.8 and 45.0 ng/mL for the test sample. In contrast, for vitamin E, there was a 95% chance that the true mean was between 17.9 and 23.9 ng/mL for the control sample and between 38.1 and 41.9 ng/mL for the test sample. Thus, there is a large degree of overlap in the means for vitamin D, but little overlap for vitamin E. If you wanted to distinguish between the two groups in the vitamin D study, you would need to carry out more experiments to reduce the standard error (see earlier).

Often in science, we measure changes in one parameter (the "dependent variable") when another parameter (the "independent variable") is changed. The resulting data may then be presented in a table or a graph. Again, it is important to establish whether the changes that are being observed are significant. As an example, let us consider an experiment where we are examining a dose–response to vitamin supplementation. A number of people (the population) are divided into different groups, and each group is fed a supplement containing a different amount of the vitamin for a month. Then, the concentration of vitamin D in their bloodstream is measured. You might want to establish how much vitamin D you need to feed someone to get a significant increase in the vitamin level in their bloodstream. As an example, we provide some (imaginary) data in Table 5.2.

**Table 5.2** Example of the use of ANOVA to determine whether there was a significant difference between consecutive data points in an experiment

| Measurements | Vitamin D dose given to test subjects in different groups (International Units) | | | | |
| --- | --- | --- | --- | --- | --- |
| | Dose 1 (0 IU) | Dose 2 (0.5 IU) | Dose 3 (1 IU) | Dose 4 (2 IU) | Dose 5 (5 IU) |
| Person 1 | 13.5 | 26.3 | 28.5 | 40.3 | 45.3 |
| Person 2 | 20.3 | 20.3 | 35.7 | 46.7 | 50.3 |
| Person 3 | 16.8 | 19.6 | 30.3 | 39.3 | 48.5 |
| Person 4 | 22.7 | 30.3 | 34.5 | 41.2 | 51.2 |
| Mean | **18.3** | **24.1** | **32.3** | **41.9** | **48.8** |
| SD | **4.0** | **5.1** | **3.4** | **3.3** | **2.6** |
| Sig Diff. ($p < 0.05$) | A | AB | B | C | C |

In this example, it is assumed that the vitamin D levels in blood collected from groups of individuals who took different doses of a vitamin supplement were tested after a month. A significant difference was defined at the 95% confidence level ($p < 0.05$). The $p$ value was calculated using single factor ANOVA (*Data Analysis* tool, Excel). Different capital letters in the last row indicate that the samples were significantly different

Initially, an ANOVA test (single factor) was carried out on the dataset shown in Table 5.2 (using the Data Analysis tool in Excel). The p-value obtained from this test was below 0.05 ($p < 0.05$), which meant that there was a significant difference between the means of at least some of the samples. As a result, a further test was carried out to establish which samples statistically differed from each other. This was performed using the Tukey's HSD (Honest Significant Difference) test. This kind of statistical analysis usually requires specialized software, which is often made available to graduate students for free or for a small fee from a university. For example, we used SPSS software to calculate the significant differences between the pairs of samples at the 95% confidence level. Samples with different letters (A, B, or C) in the last row are considered to be significantly different from each other. Thus, this analysis shows that there was only a significant change in the vitamin blood concentration when people took supplements containing 1 IU or more.

Statistics is an extremely powerful tool for research scientists, and it is impossible to cover all aspects of it here. Instead, it is strongly recommended that graduate students take a course in statistics and find a good book that clearly explains the statistical methods that are appropriate for designing and interpreting their studies (see Resources section for some examples).

## Sources of Error

Obviously, you want your experiments to give results that are as accurate as possible. It is therefore important to eliminate any potential sources of error. There are three common sources of error in experiments that lead to incorrect results being obtained:

### *Blunders*

These errors occur when the analytical procedure is not carried out correctly, *e.g.,* the wrong chemical reagents were used, some of the sample was spilt, or a measurement was recorded incorrectly. It is partly for this reason that measurements should be repeated several times using freshly prepared samples. Blunders are usually easy to identify and can be eliminated by carrying out the analytical method again more carefully. If a blunder does occur, then there are statistical methods that allow you to reject the data from your calculations of the mean and standard deviation, such as the Q-test discussed later.

### *Random Errors*

This source of error produces data that vary in a random fashion from one measurement to the next and includes factors such as:

- *Instrument fluctuations*: The electrical supply used to power many analytical instruments is inherently noisy, which can lead to random fluctuations in the results obtained.
- *Procedural fluctuations*: Researchers can make mistakes when performing their experiments that introduce random errors, such as when incorrect timing, mis-reading instrument scales, or inconsistencies in applying experimental procedures.
- *Environmental fluctuations*: Natural fluctuations in temperature, humidity, or air pressure in a laboratory could cause random fluctuations in the results obtained.
- *Sample fluctuations*: In experiments that involve sampling, random errors can arise from inherent variations within the samples themselves. For instance, if a study involves measuring the height of plants in a field, natural differences between individual plants, such as genetic variations or growth patterns, can introduce random errors.

Random errors tend to average out over repeated measurements, so that a good estimate of the true mean can still be obtained provided that sufficient measurements are made. The magnitude of the random errors determines the spread (standard deviation) of a measurement. Random errors from different sources are accumulative, *i.e.,* two sources of random error combine to make a larger error, which can be calculated using appropriate equations (see *Propagation of Errors* section).

## Systematic Errors

A systematic error produces results that consistently deviate from the true answer in a systematic way, *e.g.,* the measurements may always be 10% too high or 10% too low. There are numerous sources of systematic error, including:

- *Instrument biases*: If an analytical instrument used in an experiment is not calibrated or operating properly, it can introduce systematic errors. For instance, an improperly calibrated pipette that consistently administered a volume of fluid that was higher or lower than assumed would lead to a systematic error. Similarly, an improperly calibrated balance that consistently displayed a mass that was above or below the true value would lead to a systematic error.
- *Procedural biases*: Systematic errors can occur due to biases in the experimental procedures or techniques employed. For example, a researcher may consistently overload a balance when weighing samples, or they may consistently adjust the pH to slightly higher than the target value, which results in systematic biases in these values.
- *Environmental biases*: Rather than varying randomly from measurement to measurement, environmental conditions in the laboratory (such as temperature, humidity, or atmospheric pressure) may vary in a consistent way. For instance, the temperature and humidity of a laboratory may be higher in summer than winter, which could introduce a systematic error between measurements made at different times of the year.

- *Sampling biases*: Systematic errors can arise from biases in the sample selection process. If the sample chosen for an experiment is not representative of the entire population or if certain groups or characteristics are overrepresented or under-represented, it can introduce systematic errors in the results.
- *Personal biases*: Researchers' personal biases or preconceived notions can introduce systematic errors. These biases can influence the experimental design, data collection, analysis, and interpretation of results. Examples include confirmation bias, where researchers unintentionally favor information that confirms their preconceived beliefs, or experimenter bias, where researchers' expectations or attitudes influence the outcome of the experiment.
- *Methodological limitations*: Systematic errors can occur due to limitations or assumptions inherent in the experimental methods used. These limitations can include simplifications, approximations, or assumptions that introduce systematic biases. For instance, using a linear regression model to analyze data with a nonlinear relationship would result in systematic errors.

Researchers must identify and minimize systematic errors to ensure the accuracy and reliability of their experimental results. This can be done through careful experimental design, calibration and validation of instruments, controlling environmental conditions, employing randomization techniques, and being aware of personal biases that may influence the experimental process.

Random errors and systematic errors have different effects on the accuracy and precision of scientific measurements. The difference between accuracy and precision is highlighted schematically in Fig. 5.5. Accuracy is achieved when the

**Fig. 5.5** Schematic representation of the difference between accuracy and precision. Accuracy is when the calculated value is close to the true value (the bullseye), whereas precision is how close the measurements are to each other

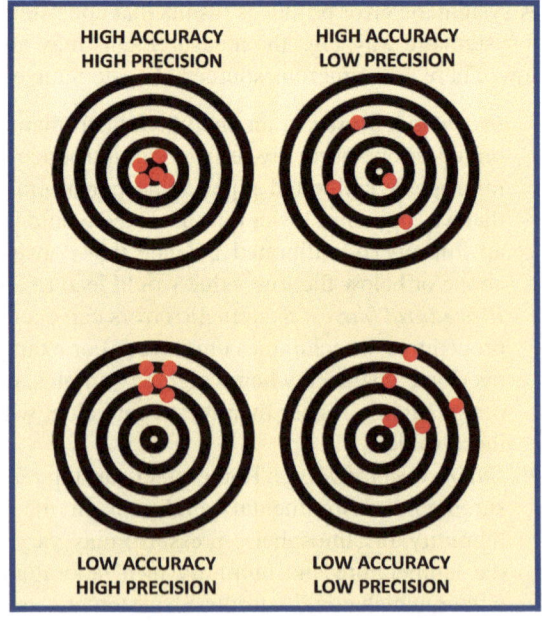

measured value is close to the true value, whereas precision is achieved when all of the results are close to each other. Ideally, you want your measurements to be both accurate and precise.

To make accurate and precise measurements, it is important when designing and setting up an analytical procedure to identify the various sources of error that may occur and to minimize their effects. Often, one particular step will be the largest source of error, and the best improvement in accuracy or precision can be achieved by minimizing the error in this step. We make this more concrete by providing a practical example from one of our PhD studies (DJM), where measurements of the velocity of ultrasonic waves in food materials were used to provide information about their composition. The ultrasonic velocity was calculated by dividing the distance the ultrasonic waves traveled through a material by their time of flight ($v = d/t$). The time of flight could be measured to be better than 0.01% using an oscilloscope, but the distance could only be measured to be approximately 1% using a micrometer. This meant that the overall accuracy of the ultrasonic velocity measurements was approximately 1%. Therefore, to improve the accuracy of these measurements, it was much more important to increase the accuracy of the distance measurements rather than the time measurements.

## Rejecting Data

When carrying out experiments, it is sometimes observed that one of the measured values is very different from all the others, which may or may not have been caused by a blunder in the analytical procedure. Statistical methods, such as the Q-test, can be used to decide whether a particular measurement can be rejected:

$$Q = \frac{X_{Bad} - X_{Next}}{X_{High} - X_{Low}} \tag{5.7}$$

Here, $X_{Bad}$ is the questionable "bad" value, $X_{Next}$ is the next closest value to $X_{Bad}$, $X_{High}$ is the highest value of the entire dataset, and $X_{Low}$ is the lowest value of the entire dataset. If the Q-value is higher than the value given in a Q-test table for the number of samples being analyzed, then it can be rejected (Table 5.3).

For example, if five measurements were carried out and one of the measurements was very different from the rest (*e.g.,* 20, 22, 25, **50**, 21), having a Q-value of 0.84, then it could be safely rejected (because it is higher than the value of 0.64 given in the Q-test table for five observations). Conversely, if one of the measurements was not very different from the rest (*e.g.,* 20, 22, 25, **29**, 21), having a Q-value of 0.50, then it could not be rejected.

**Table 5.3** Q-test table: A "bad" data point can be rejected based on statistical principles if the Q-value calculated using Eq. 5.7 is greater than the value for the appropriate number of observations. The rejection of data principle is also shown graphically

| Number of observations | Q-value for data rejection (90% confidence level) |
|---|---|
| 3 | 0.94 |
| 4 | 0.76 |
| 5 | 0.64 |
| 6 | 0.56 |
| 7 | 0.51 |
| 8 | 0.47 |
| 9 | 0.44 |
| 10 | 0.41 |

# Propagation of Errors

When calculating the final result of an experiment, it is often necessary to combine different measurements together, each with its own error. These individual errors accumulate to determine the overall error in the final result. For random errors, there are simple rules to calculate the accumulated error in the final result.

For addition (Z = X + Y) and subtraction (Z = X−Y) calculations, the overall error can be calculated from the errors in the individual measurements using the following expression:

$$\Delta Z = \sqrt{\Delta X^2 + \Delta Y^2} \tag{5.8}$$

For multiplication (Z = XY) and division (Z = X/Y), the overall error can be calculated using the following equation:

$$\Delta Z = Z \sqrt{\left(\frac{\Delta X}{X}\right)^2 + \left(\frac{\Delta Y}{Y}\right)^2} \tag{5.9}$$

Here, $\Delta X$ is the standard deviation of the mean value X, $\Delta Y$ is the standard deviation of the mean value Y, and $\Delta Z$ is the standard deviation of the calculated value Z. These simple rules are used to calculate the overall error in the final result.

As a simple example, let us assume you want to calculate the percentage protein content of a food and you previously measured the mass of the extracted protein ($M_P$) and the mass of the total food ($M_T$):

$$M_P = 3.1 \pm 0.3\,g$$

$$M_T = 10.5 \pm 0.7\,g$$

The equation used to calculate the protein content of the food is:

$$\%\text{Protein content} \quad 100 \quad M_P / M_T$$

To calculate the mean and standard deviation of the protein content, we need to use the multiplication rule (Z = X/Y) given by Eq. 5.9. Initially, we assign values to the various parameters in the appropriate propagation of error equation:

$$X = 3.1; \Delta X = 0.3$$
$$Y = 10.5; \Delta Y = 0.7$$

Z = mean protein content; $\Delta Z$ = standard deviation of protein content
Thus,

$$Z = 100 \times X / Y = 100 \times 3.1 / 10.5 = 29.5\%$$

$$\Delta Z = Z \times \left[ \left( \Delta X / X \right)^2 + \left( \Delta Y / Y \right)^2 \right]^{\frac{1}{2}} = 29.5\% \times \left[ \left( 0.3 / 3.1 \right)^2 + \left( 0.7 / 10.5 \right)^2 \right]^{\frac{1}{2}} = 3.5\%$$

Hence, the protein content of the food is 29.5 ± 3.5%. In reality, it may be necessary to carry out a number of steps in a calculation, some that involve addition/subtraction and some that involve multiplication/division, *e.g.*, (A-B)/(C-D). When carrying out this kind of multiplication/division calculation, it is necessary to complete the addition/subtraction calculations first. In this case, one would work out the errors for the (A-B) and (C-D) calculations and then use these in the division calculations.

## Significant Figures and Rounding

A common mistake in many scientific manuscripts and presentations is that too many *significant figures* are reported (Table 5.4). The number of significant figures that should be reported in a final result is determined by the standard deviation of the measurements. A good rule of thumb is to report the standard deviation to 2 significant figures and then report the mean to the same number of decimal places as the standard deviation. Some examples of this rule are given in Table 5.4. When the final number in the standard deviation or mean is 0.5, then the rule is to round upwards.

There are also rules for calculating the number of significant figures in the final result of calculations that involve combining data.

- *Multiplication/division*: For multiplication ($Z = X \times Y$) and division ($Z = X/Y$), the significant figures in the final result ($Z$) should be equal to the significant figures in the number from which it was calculated ($X$ or $Y$) that has the lowest number of significant figures. For example, 12.312 (5 significant figures) × 31.1 (3 significant figures) = 383 (3 significant figures).
- *Addition/subtraction*: For addition ($Z = X + Y$) and subtraction ($Z = X - Y$), the significant figures in the final result ($Z$) are determined by the number from

**Table 5.4** It is important to report your data to the correct number of significant figures in your calculated mean values, which depends on the standard deviation

| Sample | Incorrect | Correct |
| --- | --- | --- |
| A | 12.3452 ± 0.1254 | 12.35 ± 0.13 |
| B | 5342.20 ± 32.91 | 5342 ± 33 |
| C | 56,340.78 ± 200.23 | 56,340 ± 200[a] |
| D | 0.002910 ± 0.000567 | 0.00291 ± 0.00057 |

Report the standard deviation to 2 significant figures, and then the mean to the same number of decimal places

[a] In this case, it would be better to report the data as (56.34 ± 0.20) × $10^3$ since it is unclear whether the standard deviation is reported to an accuracy of 10 or 100

which it was calculated ($X$ or $Y$) that has the last significant figure in the highest decimal column. For example, 123.4567 (last significant figure in the "0.0001" decimal column) + 5.3 (last significant figure in the "0.1" decimal column) = 128.8 (last significant figure in the "0.1" decimal column). Or, 1310 (last significant figure in the "10" decimal column) + 12.1 (last significant figure in the "0.1" decimal column) = 1320 (last significant figure in the "10" decimal column).

When rounding numbers: always round any number with a final digit less than 5 downwards, and 5 or more upwards, *e.g.*, 23.453 becomes 23.45; 23.455 becomes 23.46; 23.458 becomes 23.46. It is common practice to carry extra digits throughout the calculations and then round off the final result.

## Standard Curves: Regression Analysis

When carrying out certain kinds of analytical procedures, it is necessary to prepare standard curves to determine the concentration of a specific substance in a sample. For example, a series of protein solutions of known protein concentration could be prepared, and then the absorbance of light at 280 nm could be measured using a UV–visible spectrophotometer. For dilute protein solutions, there is a linear relationship between the absorbance and protein concentration. Consequently, the protein concentration in a solution can be determined by measuring its absorbance, provided a suitable calibration curve has been constructed. In general, a linear relationship between the independent variable (X) and the dependent variable (Y) is described by the following equation:

$$Y = mX + c \tag{5.10}$$

Here, $m$ is the slope of the line, and $c$ is the intercept of the line with the y-axis. The values of $m$ and $c$ can be determined by finding the best fit between the experimental data and this linear equation using *regression analysis*. Regression analysis finds the line that gives the minimum deviation between the experimental measurements and the equation. How well the linear equation fits the experimental data is expressed by the coefficient of determination ($r^2$), which is the regression coefficient ($r$) squared. This coefficient of determination varies from 0 (no correlation) to 1 (perfect correlation). In the physical sciences, a coefficient of determination greater than approximately 0.5 is considered to be a good fit, but the higher the number is, the better the fit. Graphical representations of different coefficients of determination are shown in Fig. 5.6, which highlight the level of shared variance in two sets of data (the gray areas). The degree of overlap increases as the correlation between the two sets of data increases.

Examples of a good correlation and bad correlation between the dependent and independent variables are shown in Fig. 5.7. In the example given for the protein calibration curve, the protein concentration of an unknown sample could be

**Fig. 5.6** Examples of the correlation between the variance of different samples with different coefficients of determination ($R^2$) – the overlap increases as the correlation increases

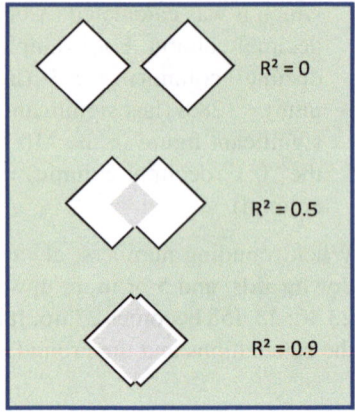

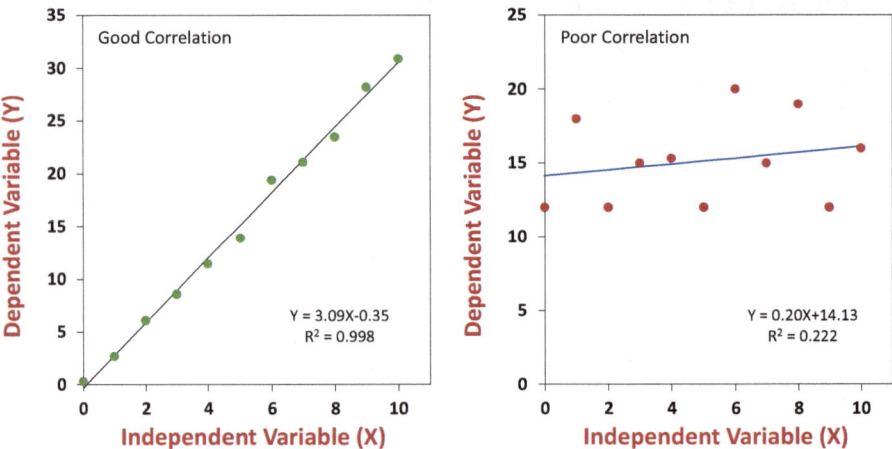

**Fig. 5.7** Examples of good and poor correlations between dependent and independent variables

determined from the equation: $X = (Y - c)/m$, where $X$ is the protein concentration, $Y$ is the measured absorbance, $c$ is the intercept, and $m$ is the slope of the linear regression analysis. These kinds of linear regression analysis can easily be carried out in spreadsheet programs (such as Excel).

**Fig. 5.8** Examples of using curve fitting methods to obtain valuable information. In this case, an exponential model is used to fit the degradation of vitamin D after heating for different times. As a result, a rate constant $k$ can be determined

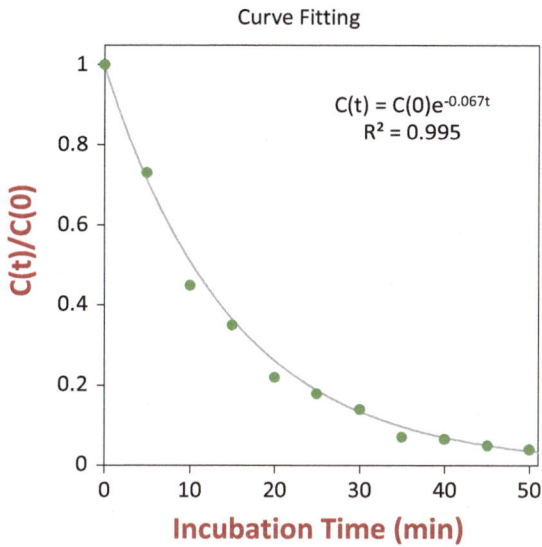

## Modeling Trends

In many studies, it is useful to use mathematical models to describe the trends observed in the data obtained from experimental measurements or empirical observations. These mathematical models can help to quantify the changes observed. For instance, the degradation of vitamin D in milk over time might be described by an exponential equation: $C(t) = C(0) \exp(-kt)$, where $C(t)$ is the vitamin concentration remaining at time $t$, $C(0)$ is the initial vitamin concentration, and $k$ is a first-order rate constant (Fig. 5.8). By fitting this equation to the experimental results, it is possible to determine the rate constant (in this case $k = -0.067$ min$^{-1}$). This value can then be used to compare the impact of different factors on the rate of vitamin D decay, such as temperature, light exposure, or pH.

Some simple equations commonly used to model experimental data are shown in Table 5.5. The most suitable equation for a particular study depends on the change in the dependent variable (Y) with the independent variable (X), *e.g.*, linear, quadratic, or exponential, to name a few common ones.

If a suitable equation can be identified, then the best fit between the equation and experimental data can be found, *e.g.*, by minimizing the square of the difference between the Y values predicted by the equation and the measured or observed Y values. The constants in the equations can then be obtained (A, B, and/or C), which can be used to quantify the observed effects. The results from experiments carried out using different treatments can then be compared. For instance, in the example given earlier, the effect of pasteurization temperature on the rate of decay ($k$) of vitamin D could be determined by measuring the change in vitamin concentration with time at different temperatures and fitting the exponential model to the data.

**Table 5.5** Examples of a few simple equations commonly used to model experimental data

| Model | Equation | Comments |
|-------|----------|----------|
| Linear | $Y = A + BX$ | Linear<br><br>Y increases or decreases linearly with X |
| Quadratic | $Y = A + BX + CX^2$ | Quadratic<br><br>Y increases or decreases quadratically with X |

(continued)

**Table 5.5** (continued)

| Model | Equation | Comments |
|---|---|---|
| Exponential | $Y = A \exp(BX)$ | <br>Y increases or decreases exponentially with X |

By fitting these models to data, one can obtain information about the constants (A, B, and/or C), which can be used for quantifying changes

## Correlation and Causation

It is important when interpreting any trends in your experimental data that you do not confuse correlation with causation. Correlation refers to relationships between two variables that change together in a predictable way. For example, a study might find that there is an increase in lifespan as the amount of salmon in people's diet increases. This correlation could be because salmon is a good source of protein and omega-3 fatty acids, but it could also be because people who can afford to buy salmon are wealthier. As a result, they have a healthier lifestyle and can access better health care, which allows them to live longer. Thus, a correlation might be due to causation, but it might also be due to something else. To establish causation, one must find a cause-and-effect relationship between the two variables being studied. Ideally, there should be a good physical, chemical, or biological reason for the observed effect. For example, if you discovered an increase in the vitamin D concentration in people's bloodstreams after eating vitamin supplements, then there is a strong reason to believe that this correlation is due to causation. There are highly plausible explanations that consuming more vitamin D in your diet will lead to a vitamin D increase in your bloodstream. This oil-soluble vitamin is known to be absorbed in our gastrointestinal tracts and accumulates in our bodies.

In summary, correlation refers to the degree to which two variables are related, while causation refers to the fact that one variable directly causes a change in the other variable. Thus, just because two variables are correlated does not necessarily mean that one causes the other – you would have to do further research to establish a plausible cause-and-effect mechanism.

## Establishing Causation: Identifying the Origin of Observed Effects

The impact of your research will be greatly increased if you can establish the cause of an observed phenomenon in terms of fundamental physical, chemical, biological, or sociological principles. Establishing causation can take your research from simply observing and reporting interesting or important phenomena to providing fundamental insights and understanding. In some cases, you may be able to use a mathematical or computational model based on fundamental scientific principles to describe your results and make predictions that can be tested.

One of the things I (DJM) am most proud of from my own graduate studies was finding a theoretical model that could be used to model my experimental results, provide a fundamental physical interpretation of what I was observing, and allow me to make predictions of what factors should impact my experiments. My PhD research was on characterizing emulsions using ultrasonic spectroscopy. These emulsions consisted of tiny oil droplets that were suspended in water. My experiments showed that the velocity and attenuation of ultrasonic waves passing through the emulsions depended on the size of the oil droplets they contained, but I had no idea why. My professor gave me a scientific paper that had been published in the 1950s where the researchers had developed a mathematical model to describe the propagation of ultrasonic waves through emulsions. This paper consisted of pages and pages of complex mathematical equations that had been derived by describing the interactions of high-frequency pressure waves with small particles. I was not very familiar with mathematics or computers, so I spent a few months learning math and programming. Eventually, I was able to program the theory and use it to interpret my results. I was very excited when the predictions of the theory agreed very closely with my experimental data. The theoretical model enabled me to identify why the velocity and attenuation of ultrasonic waves changed with the size of the oil droplets in terms of basic physical principles (viscous and thermal losses of energy at the droplet surfaces). The model also predicted that the propagation of ultrasonic waves through emulsions depends on the oil droplet size, concentration, and composition, which helped me design new experiments. My research study went from simply describing a phenomenon to truly understanding it and being able to make predictions. This allowed me to publish my research findings in high impact physics and physical chemistry journals, rather than the food science journals I normally published in.

Sometimes, the phenomena that you are observing in your studies may be too complex to understand with existing physical theories. However, we would certainly recommend that you try your best to find the latest scientific theories to model, understand, and predict your findings, as this will greatly increase the impact of your research. The different general approaches for interpreting experimental results in the physical sciences were outlined in an earlier chapter, including black-box approaches that use simple correlation and artificial intelligence and gray- and white-box approaches that use physicochemical concepts, theoretical modeling, and computer simulations. For this reason, we only briefly highlight them here, giving examples from some of our research on understanding the properties of meat substitutes assembled from plant proteins and polysaccharides:

- *Black-box – Correlation*: In this approach, a researcher simply tries to establish a correlation between two or more variables without providing any insights into the fundamental reason why this correlation exists. For instance, a researcher may simply report data (figures or tables) that show the color or texture of plant-based meat analogs changes as the pH and salt concentration changes, without giving any reasons why.

- *Black-box – AI*: In this approach, artificial intelligence (AI) is used to establish correlations between input and output variables. Again, no insights into the fundamental origin of the observed phenomenon are obtained, but the AI model generated may have good predictive power. For instance, a researcher may create a wide range of plant-based meat substitutes using different ingredients and processing methods (inputs) and then measure their appearance, texture, and stability (outputs). The AI model may then be able to identify the best combination of ingredients and processing methods needed to obtain a meat substitute that is similar to real chicken.

- *Gray-box – Descriptive*: In this approach, a researcher may carry out experiments that show a correlation between two variables but then interpret the observed changes in terms of fundamental physicochemical or biological phenomena. For instance, a researcher may hypothesize that the dependence of the appearance and texture of plant-based substitutes on pH and ionic strength is due to changes in the electrical charge on the proteins and polysaccharides, which influences the electrostatic attraction and repulsion between them. This approach is often taken when the system being studied is so complex that it is difficult to describe it with the current level of theoretical or computational tools.

- *White-box – Theoretical*: In this approach, a researcher may carry out experiments that show a correlation between two variables and then interpret the results using a mathematical model based on the fundamental science of the phenomenon involved. This could involve identifying an existing theory in the literature, modifying an existing theory, or developing an entirely new theory. For instance, a researcher trying to understand the properties of plant-based meat substitutes might use a theoretical model that was originally derived to predict the mechanical properties of composite polymers, based on the number, type, and interactions of polymer molecules present. It is therefore important to look beyond your own research field when trying to find a suitable theory to interpret your results.

- *White-box – Computational*: Recently, there have been major advances in computational science that have led to computer programs capable of modeling the physical and chemical phenomena occurring within materials and how they impact their overall physicochemical properties. A researcher may therefore try to understand what is happening in their experiments by using these computer programs to model the behavior of the different kinds of molecules involved (**Fig. 2.12**). These models can provide fundamental insights into the importance of different properties of the system on their behavior, such as the number, size, shape, and interactions of different molecules.

To describe and interpret your research findings, it is important to carry out an extensive literature search to better understand the fundamental science behind the phenomenon you are studying. Here, we have mainly focused on the physical sciences, but similar approaches are also important in the biological and sociological sciences.

## Plotting Data: Which Type of Graph or Table to Use?

After you have collected all the results from your experiments and analyzed them (for example, calculating means, standard deviations, and statistical differences), then you need to present them in a suitable format. There are numerous options for presenting data, including various tabular or graphical formats, which each have advantages and disadvantages. There are several factors to consider when selecting the most appropriate format. The most important is that your data is presented in a clear and concise fashion that is easy to understand and that highlights the importance of your findings. If you do use a graphical format, then it is important that the graphs are prepared to be visually appealing (see Chap. 6). Here, we highlight some of the most important ways of presenting scientific data and highlight some of the advantages and disadvantages.

### *Tables*

Tables are useful when you have a lot of different treatments in your experiments, and it would be confusing to plot them all in a single graph. They are also useful when you want to provide the numerical values of your experiments to other scientists, who could then use them in their own research (which may increase the impact and citations of your manuscript). However, large lists of numbers in a table are not as visually appealing or intuitively informative as plotting data in a graphical format. It is much easier to see trends in data when it is plotted as a graph, rather than presented in a table. Ideally, it would be advantageous to present your data in both tabular and graphical forms, but most journals do not allow this due to limitations of space. However, some journals now encourage authors to include additional data in a *Supplementary Material* section that is not published with the main manuscript but can be accessed online. It is therefore possible to represent the same data as a graph in the main article to improve its comprehension, but as a table in the *Supplementary Material* section so that other people have access to the quantitative values. It is important to report data to the correct number of significant figures when reporting data in a table (see earlier).

## Bar Charts

Bar charts are useful when you want to show changes in your data as a function of some variable (like time or dose), compare data in different categories (like before and after heating), and make a strong visual impact. They are often used when the x-axis variable is in the form of categories (such as "group 1", "group 2", *etc.*) rather than a continuous scale (such as time or temperature). For example, you may want to compare the hardness of four polymer formulations with different compositions ("Polymer 1", "Polymer 2", *etc.*) before and after heating (Fig. 5.9).

## Pie Charts

Pie charts are useful for visually representing proportions or percentages of a whole. For instance, you may want to show the percentage of people who are vegans, vegetarians, flexitarians, and omnivores in the United States, or you may want to show the regions of the world where the most meat is consumed as a proportion of the global population (Fig. 5.10). Pie charts are usually better when there is only a limited number of categories to show; otherwise, they become difficult to read and comprehend. One of the main advantages of pie charts is that they can be visually appealing and can easily highlight differences between categories.

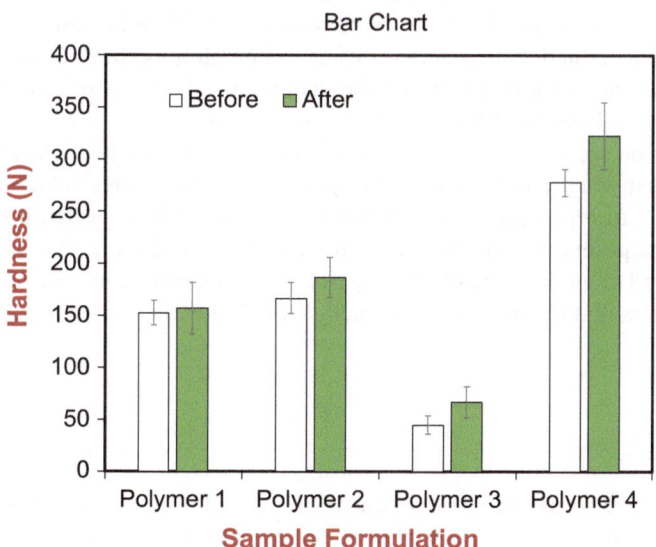

**Fig. 5.9** Plotting data as a bar chart is useful for showing trends or comparing samples in a visually appealing way

**Fig. 5.10** Plotting data as a pie chart is often useful when you want to show the proportion or percentage of a whole in a visually appealing way. In this case, the proportion of meat consumption in different continents in 2018 is shown, which highlights that Asia and America are the largest consumers of meat

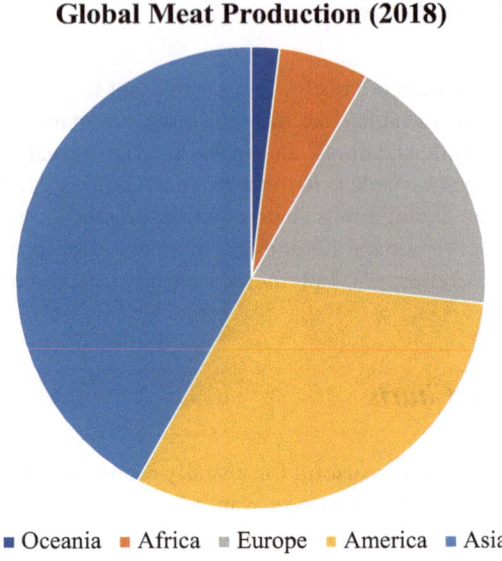

**Global Meat Production (2018)**

■ Oceania  ■ Africa  ■ Europe  ■ America  ■ Asia

## Scatter Plots

Scatter plots are used to plot a dependent variable (y-axis) as a function of an independent variable (x-axis) when the independent variable changes in a continuous manner, such as time, temperature, or concentration. They are useful for showing trends in data and for comparing data from different treatments. It is important not to include too much data in scatter plots because they then become confusing and difficult to understand. One issue to consider when plotting the data on scatter plots is whether to include a line between the data points. The data points represent the actual data that have been measured or observed. If you plot a line between each point, then you are assuming that the data is behaving in a certain way at intermediate points that you did not measure. Just plotting the data points is therefore a more accurate way of representing your experimental data, but it is often useful to connect the data points with a smooth or straight line to make the plot more visually appealing and to guide the reader to any trends. In this case, it may be useful to state in your figures that the lines are only included to guide the reader (Fig. 5.11).

## Line Plots

Line plots are also used to plot a dependent variable (y-axis) as a function of an independent variable (x-axis), so they look very similar to scatter plots. However, the way the data are plotted on the x-axis is different. For a line plot, each successive data point is equally spaced along the x-axis, regardless of its value. Line plots are therefore most useful for representing numeric or nonnumeric x-values with equal intervals between them, such as numbers (1, 2, 3, …), months (Jan, Feb, Mar, …), or years (1990, 2000, 2010, …). As an example, a line plot representing the change

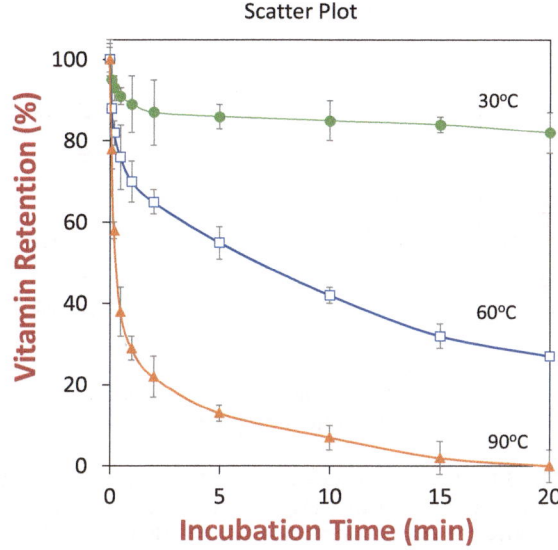

**Fig. 5.11** Scatter plots are useful when you want to show trends in your data or when you want to compare data from different treatments. This plot shows (made up) data for the impact of storage temperature on vitamin degradation over time

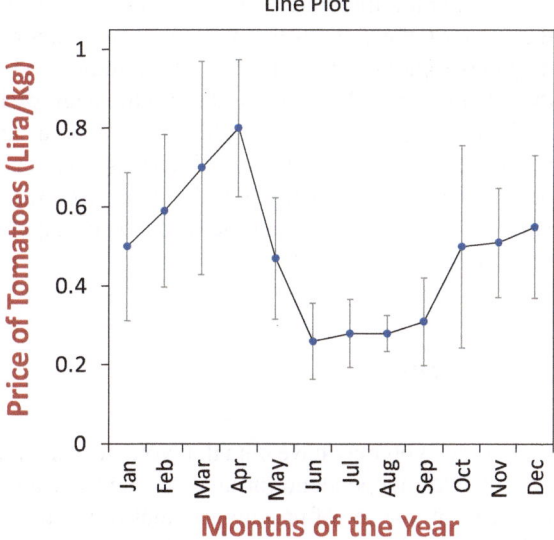

**Fig. 5.12** The average price of tomatoes per kilogram in Turkey throughout the year from 2000 to 2010 (prices in Turkish Lira) shown as a line plot

in the price of tomatoes throughout the year is shown in Fig. 5.12, which shows that they are cheaper in the summer months when they are most abundant.

One must be careful when using line plots to represent certain kinds of data, especially when there are not equal intervals between the x-axis data points. As a concrete example, the same data are plotted using scatter and line plots in Fig. 5.13. In this example, the change in vitamin concentration with incubation time is plotted when a food is held at a high temperature where the vitamin tends to degrade. For

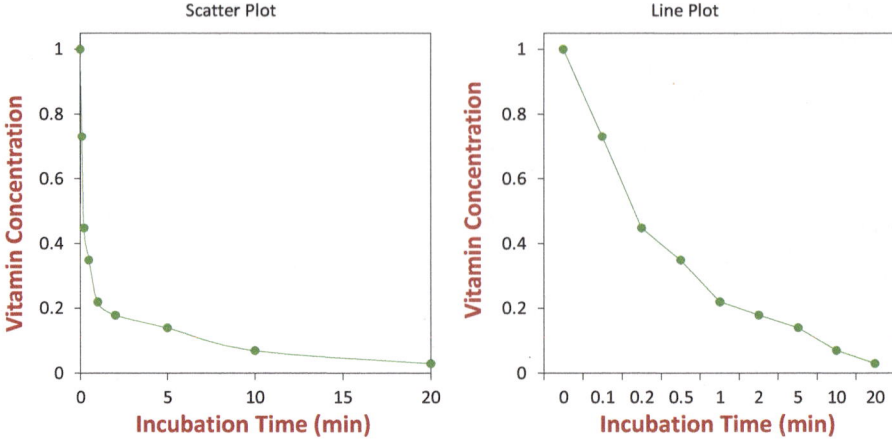

**Fig. 5.13** Comparison of scatter and line plots for representing the same data, where the x-axis value may vary continuously. In this case, the change in vitamin concentration over time is reported. The scatter plot more clearly shows that the vitamin rapidly degrades in the first few minutes

the scatter plot, there are equal times between each tick mark on the x-axis, but for the line plot, there are different times. The scatter plot clearly shows that the vitamin degrades rapidly during the first few minutes and then more slowly later, whereas the line plot provides less insight into the degradation rate of the vitamins.

Generally, a line chart should primarily be used when your data has numeric or nonnumeric x-values with equal intervals between them (such as numbers, months, or years), whereas a scatter plot should be used when the x-values are numerical and vary continuously, such as time (seconds or days), temperature (°C), or concentration (g/mL).

## Surface Plots

Sometimes you may have data that you measure a dependent variable (Y) as a function of two independent variables (X and Z). For instance, you may measure the hardness of a series of protein gels that were prepared using a range of different pH values and salt concentrations. In this case, you may want to use a surface plot to represent the data (Fig. 5.14). Surface plots can provide a visually appealing representation of trends in the data. However, they can sometimes obscure part of the data, making it difficult to interpret, so they should be used with caution.

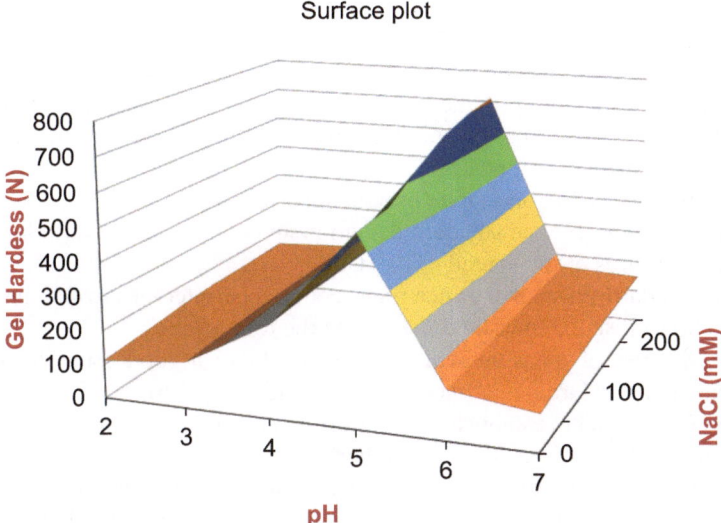

**Fig. 5.14** Surface plots can be useful for highlighting trends when a dependent variable changes with two independent variables. In this case, the impact of pH and NaCl concentration on the hardness of protein gels is shown (made up data)

## Concluding Remarks

Scientific studies typically generate a large amount of data that needs to be analyzed, interpreted, and communicated. The values of scientific measurements and observations vary due to the natural variability of the world. A good understanding of statistics is therefore critical to the proper design and interpretation of most scientific studies. The proper use of statistics leads to results that are more accurate and precise. We would therefore recommend that you take a basic statistics course or at least become familiar with the most important statistical concepts relevant to your field of study. You should be aware that your results are always prone to experimental errors for many different reasons, including blunders, random errors, and systematic errors. It is important to carefully consider all the possible sources of errors in your studies and eliminate or reduce them. Representing your scientific findings in a clear and concise manner that is visually appealing to readers will also increase the impact of your research. Wherever possible, we strongly recommend that you try to interpret your results using fundamental scientific principles (such as physics, chemistry, and biology), rather than simply developing correlations (black-box approach). The black-box approach produces new knowledge about the world, but the fundamental approach creates new understanding.

**Key Points**
- Descriptive and inferential statistics are used by research scientists. Descriptive statistics provide estimates of the nature of a dataset. Inferential statistics infer information about an entire population from a smaller subset of samples.
- The mean, median, and mode provide measures of the central tendency of a dataset, whereas the standard deviation and coefficient of variation provide measures of the spread of the data.
- The standard error of the mean and confidence interval provide insights into how close the measured mean is to the true mean.
- The number of measurements needed to obtain a good estimate of the mean and standard deviation can be estimated (Eq. 5.6).
- ANOVA is used to establish whether the means of two or more samples are significantly different from each other.
- There are various sources of error in experimental measurements, including blunders, random errors, and systematic errors, which should be identified and minimized.
- Statistical tests, such as the Q test, are used to determine when apparently erroneous results can be rejected.
- The overall error in a final value can be calculated using propagation of error equations.
- Results should always be reported to the correct number of significant figures.
- It is important to distinguish between correlation and causation. Correlation refers to the degree to which two variables are related, while causation refers to the fact that one variable directly causes a change in the other variable. Just because two variables are correlated does not necessarily mean that one causes the other.
- White-, gray-, and black-box methods may be used to interpret experimental results.
- The most effective method of presenting your data should be selected, such as tables, bar charts, pie charts, scatter plots, line plots, and surface plots.

## Resources

Frost J (2019) Introduction to statistics: an intuitive guide for analyzing data and unlocking discoveries. Jim Publishing, State Collage
Nielsen SS (2017) Food analysis, 5th edn. Aspen Publication, Gaithersberg
Urdan TC (2022) Statistics in plain English, 5th edn. Routledge, New York

# Chapter 6
# Writing Your Manuscript: The Nuts and Bolts

## Getting Started

As stressed in previous chapters, writing and publishing scientific manuscripts is one of the most important things you do as an academic because it is the main way that other people find out about your work. Moreover, it helps to build your reputation in your field of study, thereby increasing your future chances of success. Writing

D. J. McClements et al., *How to be a Successful Scientist*,
https://doi.org/10.1007/978-3-031-51402-9_6

your first scientific manuscript, however, can be a daunting process, but it does get easier the more you do it. In this chapter, we highlight a general approach that can help you write good scientific manuscripts efficiently and effectively. This approach divides the writing process into several steps to make it more manageable. Some of the information presented is based on advice given by a major publisher of scientific manuscripts (Springer, 2023), but much of it comes from the authors' experience in writing manuscripts.

## *Organizing Your Data*

After you complete all your experiments, you should collect, organize, and analyze your data. As part of this process, you should compile your data into a well-organized spreadsheet file (like Excel), with each worksheet representing a single figure or table that will appear in the final manuscript. An example of a spreadsheet file used for preparing one of our own manuscripts is shown in Fig. 6.1. This file contains all of the data used to write a scientific manuscript on the ability of natural emulsifiers to form and stabilize emulsions.[1] Each worksheet in the file is given an informative label, such as "Fig 1 (PSD-emulsifier)", which includes all the data for Fig. 1 of the

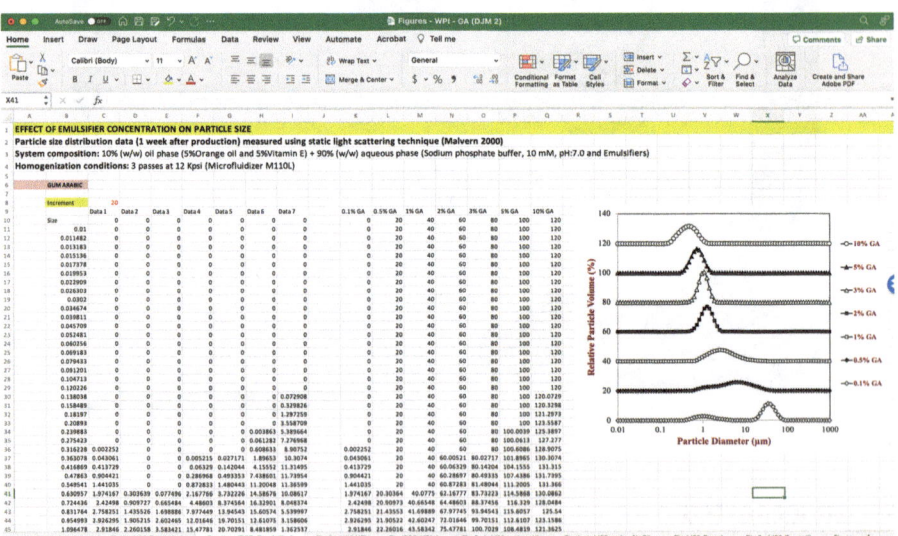

**Fig. 6.1** Example of an Excel file containing the data used to prepare a scientific manuscript on the ability of natural emulsifiers to form and stabilize emulsions. Each worksheet in the file contains data for one figure and has a name that reflects its contents

---

[1] An emulsion consists of tiny droplets of one liquid that are dispersed in another immiscible liquid, like the tiny droplets of oil dispersed in water in food products like milk, creams, or salad dressings.

manuscript on the impact of emulsifier concentration on the particle size distribution (PSD) of the emulsions.

Initially, the figures can be organized according to the plan you made for the different subsections of your *Results and Discussion* section before you started your experiments (see Chap. 3). However, this can be modified if you think there is a more logical and engaging way of presenting your data after you complete your study. At the end of this process, you should have a single Excel file organized in a logical fashion that contains all the figures and tables that will appear in the final manuscript. You may also want to have another file containing all the schematic diagrams and photographs that will appear in your final manuscript (since it is not convenient to include these in a spreadsheet program like Excel). In this case, you can use a presentation program (such as PowerPoint) that is more convenient for drawing schematic diagrams and labeling photographs. An example of a PowerPoint file used to prepare one of our own scientific manuscripts is shown in Fig. 6.2.

## *Naming Your Folders and Files*

It is important to have well-organized folders and files on your computer hard drive, as well as a cloud database, so you can easily find your data and manuscripts. When writing a scientific manuscript, it is common to have three different kinds of files: a spreadsheet file (like Excel) containing all the data; a presentation file (like PowerPoint) containing any photographs and diagrams; and a text file (like Word) containing the written parts of the manuscript. These files are usually placed in a single folder that has a simple name that reflects its contents, such as "PB Meat – Protein type (2023)," for a folder containing the information of a study that examines the impact of protein

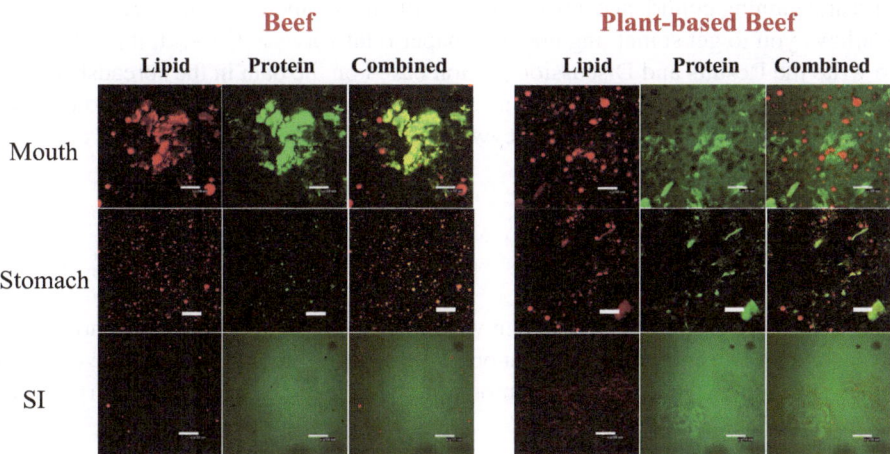

**Fig. 6.2** Example of a page from a PowerPoint file containing images used to write a scientific manuscript that compared the digestion of plant-based and real beef burgers

type on the formation of plant-based meat analogs. Each file in this folder is given a consistent name that reflects the kind of information it contains, such as "Data – PB meat protein (DJM 1)," "Figures – PB meat protein (DJM 1)," and "Text – PB meat protein (DJM 1)" for spreadsheet, presentation, and text files, respectively. The "DJM" is for the initials of the author (*e.g.,* David Julian McClements), and the number "1" is the latest version of the file that is being worked on. If substantial changes are made to a file, but you still want to keep the older version because it contains some useful information or you are not sure about the changes you made, then you can rename it "DJM 2." If somebody else works on the manuscript, then their initials can be added, *e.g.,* "Text – PB meat protein (DJM 1-IFM 1)." This helps to keep track of which version of the manuscript is the most current.

## *Organizing the Writing of Your Manuscript*

Once you have prepared the spreadsheet and presentation files containing your figures, tables, and images, you are now ready to write your scientific manuscript. You must then decide which part of the manuscript to write first. Typically, published manuscripts should be organized into a specific sequence of sections, such as title, abstract, keywords, introduction, materials and methods, results and discussion, conclusions, and references[2] (Fig. 6.3 **left**). However, when you are writing your manuscript it is often better to use a different sequence. Different people have different preferences, but a commonly used writing sequence is shown in Fig. 6.3 **(right)**. Typically, it is advisable to write the Materials and Methods section first because this describes all the reagents, analytical techniques, experimental protocols, and statistical methods used in the study, which does not require any data analysis, interpretation, or discussion. This can also be a great way of making progress and gaining confidence when you are still unsure about writing manuscripts, as it allows you to get something down on paper relatively easily. Next, it is advisable to write the Results and Discussion section based on the data in the spreadsheet and presentation files discussed in the previous section. Then, the Introduction, Conclusions, Abstract, Title, and Keywords can be written (usually in this order).

## Authors

An important decision to make when writing a scientific manuscript is who should be included as an author, and in what order the authors should be listed. The position of an author in the list impacts the amount of credit they get for the manuscript, with

---

[2] This classical structure of the body of scientific manuscripts is often given the acronym: IMRaD, which refers to the *I*ntroduction, *M*aterials and Methods, *R*esults, and *D*iscussion and Conclusions. This format may change somewhat from journal to journal and so it is always useful to check the instructions to authors.

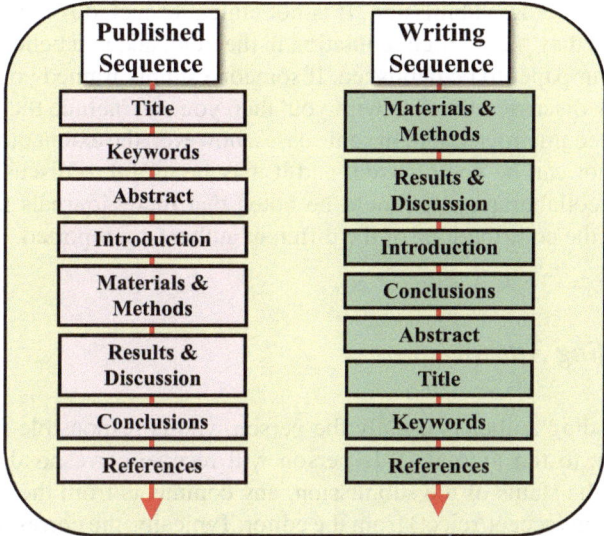

**Fig. 6.3** Scientific manuscripts typically contain several elements that are organized in a particular sequence in the final published article (left). However, when you are writing the manuscript, it is often better to write the elements in a different sequence (right) because the information presented in some sections (such as the Abstract) depends on information that is presented in other sections (such as the Results and Discussion)

the first and last positions usually being given the most credit. Several people may contribute to designing, performing, and carrying out the experiments, as well as interpreting and presenting the data and writing and editing the manuscript. In this section, we briefly discuss how to decide who should be included as an author and where they should appear in the author list.

## *What Qualifies as Authorship?*

You should only include those people who have made a significant contribution to your manuscript in the author list. A person may have contributed to the manuscript in various ways:

- They may have come up with the original idea the manuscript is based on, secured the funding that supported the research, and helped to design the research plan.
- They may have been involved in collecting, analyzing, and reporting the data.
- They may have been involved in writing and revising the manuscript.

Each author should have made an important contribution to the final manuscript and should be responsible for ensuring its accuracy and integrity, as well as approving the final version before submission. It is not ethical to include "honorary" authors simply because they have a high reputation in the field, and you believe it will make it easier for your paper to be published. If someone simply trained you on an instrument or briefly discussed the data with you, then you can include their contribution in the acknowledgments section. In some cases, however, the extent of work required to be a coauthor can be complicated, and it may be useful to discuss it with your supervisor or collaborators. It should be noted that many journals now include a section where the contributions of the different authors are reported.

## Corresponding Author

The corresponding author is usually the person who is responsible for submitting the manuscript to the journal. This person will receive correspondence from the journal about the status of the submission, any comments from the reviewers, and the final decision (accept/reject) from the editor. Typically, the corresponding author is the principal investigator of the study. This is often the professor who came up with the initial idea for the research and/or who supervised the students and post-docs who carried out the research. Occasionally, a professor may have a postdoc or senior graduate student act as the corresponding author if they make a major contribution (as this may be good for their future career). The corresponding author may go in different positions in the author list. It is also possible to have more than one corresponding author on a manuscript by including a statement like "*These people are both corresponding authors" at the end of the title page.

## Who Goes Where?

Once you have decided who should be included as an author, then it is necessary to decide who goes where in the list. Sometimes this is simple to decide because there are only a few people involved in the study and the relative importance of their contributions is clear. For instance, if there are only two or three people involved in the research (including the professor) and one of the people did most of the work, then that person usually goes first, the professor goes last, and any other person goes in the middle. However, in other cases, it can become quite contentious because two or more people may feel they have made the major contribution and should go first. In this case, it may be up to the main author to decide who should go first, usually after discussing it with the other authors and giving a clear rationale for their decision. Here, we describe some of the factors that impact the best way to order the authors:

- *First author*: The person whose name appears first in an author list is often given the most credit for doing the research reported in the manuscript, which is especially important for the career development of graduate students and postdocs.

Typically, the first author is the person who has contributed most significantly to the research. They are usually the person who has had the greatest role in designing, performing, and analyzing the experiments, as well as possibly writing the first draft of the manuscript. Being the first author is important because when there are several authors, a manuscript may get cited using only the first author's name, such as "McClements et al. (2023)." It is therefore a highly coveted position. In some cases, it is possible to designate more than one person as the first author (by putting stars next to the authors' names and including a statement such as "*These authors contributed equally to this manuscript" at the end of the title page.

- *Last author*: The person whose name appears last in the author list is usually the principal investigator who came up with the initial idea and/or who supervised the research team. This person also receives much of the credit for the manuscript and is usually the corresponding author.
- *Intermediate authors*: Following the first author, any additional authors are typically listed based on their relative contributions to the study from the most to the least. Again, if more than one person made an equal contribution, this can be highlighted by including a statement "These people contributed equally to this study." Listing the positions of the intermediate authors can be challenging because it is difficult to precisely establish the relative importance of their contributions. In this case, the main author may have to decide on the author list. They may be able to do this by compiling a list of the contributions made by the different authors and assessing whose were the most important. Alternatively, the main author may simply list the intermediate authors in alphabetical author to avoid any conflicts.

It is often advisable to decide who qualifies as an author and their relative roles at the beginning of a study. This may change as the study progresses if someone makes a larger or smaller contribution than expected. It may then be important to discuss authorship with the different people involved with the project regularly. This will help to reduce the chances of any contentious issues arising when the manuscript is being prepared for submission.

# Title

Writing a good title is important for ensuring that your manuscript gets the attention of potential readers. It is the hook that will be the first/only thing that many people use to decide whether they want to read your article. It should therefore accurately and concisely describe the subject matter of your manuscript and convey the main findings. It should also be carefully designed to attract readers and increase the visibility (and potentially citations) of your work. One strategy for writing a good title is to make a list of the main topics covered in the study and then write a title that includes them. It is a good idea to put yourself in the mind of a potential reader and to think of what keywords they may enter into a search engine to find papers on your topic. The title should not be too long because it may then be difficult for other scientists to quickly read and comprehend its content when they are scanning through the results of a literature search that may contain a list of tens or hundreds of articles. Conversely, it should not be too short because then it may not contain sufficient information for a reader to appreciate the nature of the study. It is also important not to write a title that is full of hyperbole, like "Curcumin: The Miracle Cure for All Diseases." It is challenging to write a title that is concise, informative, and engaging, but it is certainly worthwhile putting in the effort to do this, as it can greatly influence the impact of your work.

As an example, we provide some potential titles for a manuscript on the effects of temperature and salt on the interactions between a milk protein ($\beta$-lactoglobulin) and a plant polysaccharide (pectin), as well as some comments on their suitability:

"Impact of temperature and salt on $\beta$-lactoglobulin and pectin mixtures"

– This title is too short and does not contain enough details about the purpose or importance of the study.

"Characterization of the effect of temperature and salt on the interactions between $\beta$-lactoglobulin and pectin characterized using dynamic light scattering, turbidity, and microelectrophoresis measurements"

– This title is too long. It contains a lot of information about the techniques used but does not provide any information about the purpose or importance of the study.

"Simple method for biopolymer nanoparticle fabrication: Electrostatic complexation of heat-treated $\beta$-lactoglobulin-pectin mixtures"

– This title is a better length and contains more insights into the purpose and importance of the research. In addition to containing information about the components used ($\beta$-lactoglobulin and pectin), it also contains some "hot topic" phrases that may attract readers, such as "biopolymers," "nanoparticles," and "electrostatic complexation."

The use of a colon is often desirable when designing manuscript titles as it allows you to provide a broad statement about the general area of interest, and then provide specific details about the particular topic of the study.

When designing your title, it is important to consider why your research findings might be of interest to other scientists. You should think about the titles of other people's manuscripts you have liked and what draws you to want to read them. A few additional tips about writing a good title are included here, which have been adapted from a website that provides support for technical writers (www.redwood-ink.com):

• *Use a descriptive or conclusive title:* Typically, the titles of scientific manuscripts are either descriptive or conclusive. Descriptive titles state the main focus of the study, whereas conclusive titles state the main conclusion of the study. For example, a descriptive title may be "Development of soy protein-based nanoparticles to encapsulate and deliver curcumin", whereas a conclusive title may be "Encapsulation of curcumin in soy protein-based nanoparticles increases its stability and bioavailability." Typically, a conclusive title is more informative and engaging.

- *Start the title with the most impactful words:* Potential readers typically find articles by using search engines that produce a list of different manuscripts in the area of interest. They may then rapidly scan through this list to identify the most relevant manuscripts. It is therefore a good idea to place the most important and impactful words at the beginning of the title. For example, the title of a review article "Nanoparticle-based delivery systems for application in foods" may be more impactful than "Potential applications of nanoparticle-based delivery systems in foods."
- *Include critical terms in the title:* The impact of your title can be increased by including critical terms that help readers determine the topic of your study, such as key materials, methods, or approaches. These critical terms may also be picked up by search engines, which will help readers find your article. As an example, a manuscript title may be "Fabrication of core-shell biopolymer nanoparticles from zein and pectin using antisolvent precipitation."
- *Write your title using plain language:* Your title should be written in plain language that is unambiguous and simple to understand. This will not only help other scientists to identify the importance of your work but may also attract the attention of policymakers or journalists, thereby increasing its impact. As an example, "Nanobiomimetic methodological approach for augmentation of techno-functionality of submicron polypeptide-layered triacylglycerol spherical particles" is written in a language that may be unclear to many people. A simpler, more informative, and more impactful title for this study would be "Improving the stability and performance of plant protein-coated lipid droplets using biomimetic polysaccharide coatings."
- *Be precise and concise:* Your title should be as short as possible, while still conveying the nature of the study and attracting readers. Eliminate all unnecessary words from the title of your manuscript. For instance, a title such as "A study of the influence of electrostatic interactions on the aggregation of plant proteins dispersed in aqueous solutions" (18 words) could be rewritten as "Influence of electrostatic interactions on plant protein aggregation in aqueous solutions" (10 words), without losing any meaning. Typically, your title should be approximately 10–14 words.
- *Avoid abbreviations:* Your title should not contain abbreviations that are unfamiliar to many people, who may be specialists or nonspecialists. For instance, a title like "Influence of encapsulation of HPO in PLA-nanoparticles on its SFC" would be confusing to most readers. Instead, it would be better to spell out the various acronyms so that readers are clear about the nature and purpose of the study: "Influence of encapsulation of hydrogenated palm oil in poly(lactic acid)-nanoparticles on its solid fat content." Having said this, it is appropriate to use abbreviations that are in common use and familiar to most people, like DNA

instead of deoxyribonucleic acid or AIDS instead of Acquired Immune Deficiency Syndrome.

We have taken a lot of time considering how to design a good title. However, this is time well spent because it can greatly increase the impact of your manuscript by attracting more readers.

## Abstract

A well-written abstract is another important element that can help increase the number of people who read your manuscript. As mentioned earlier, people typically find your research articles using internet search engines, such as Web of Science, Pub Med, Scopus, or Google Scholar. They typically use a few keywords or author names to identify important articles in the area they are interested in. This will result in a list of different articles that might be of interest. If someone finds the title of your article interesting, then they may read the abstract to see if it contains the kind of information they are looking for. In some cases, they may be writing an original research article where they want to use your findings to design or interpret their own experiments, which may lead to your article being cited in their manuscript. In other cases, they may be writing a review article and want to summarize your research in their manuscript. Busy researchers may rely completely on what you have written in the abstract, rather than reading the full paper. It is therefore critical that your abstract contains a concise but accurate and informative overview of your study, including the purpose of the work, the methods you used, the results you found, your interpretation of the results, and the importance of any new findings. Many journals have a strict maximum word count for the abstract (such as 150 or 250 words), which means you must fit all of this information into a relatively small space.

Another reason for writing a good abstract is that it can speed up the review of your manuscript. Journal editors often send only the title and abstract of a manuscript to potential reviewers to see if the subject matter is in their area of expertise and if they are willing to review it. Consequently, writing a clear, concise, and informative abstract can help reviewers quickly decide whether they are qualified to review your paper or not, which can shorten the time it takes to review your manuscript and get it published. If your abstract is written badly, they may not want to devote the time to plowing through the rest of the manuscript.

When preparing an abstract for a scientific manuscript, which is often done after the rest of the manuscript has been completed (Fig. 6.3, **right**), it is advisable to use a standardized structure because this leads to a more efficient and effective writing process (Fig. 6.4):

- *Context*: Typically, it is a good idea to begin an abstract with a brief statement about the general topic or problem that the study addresses. This helps the reader quickly understand the overall context and importance of the research. For instance, if you were writing a paper on the isolation and characterization of a

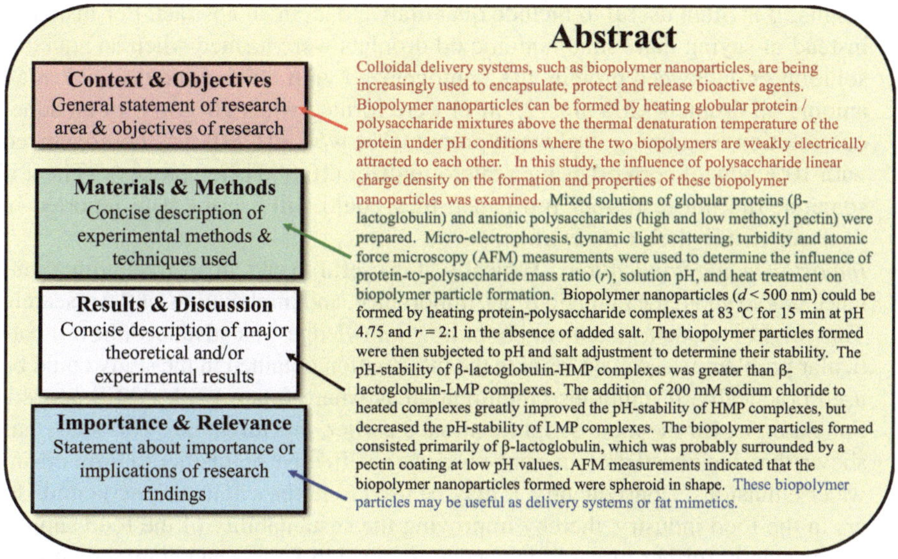

# Abstract

**Context & Objectives**
General statement of research area & objectives of research

**Materials & Methods**
Concise description of experimental methods & techniques used

**Results & Discussion**
Concise description of major theoretical and/or experimental results

**Importance & Relevance**
Statement about importance or implications of research findings

Colloidal delivery systems, such as biopolymer nanoparticles, are being increasingly used to encapsulate, protect and release bioactive agents. Biopolymer nanoparticles can be formed by heating globular protein / polysaccharide mixtures above the thermal denaturation temperature of the protein under pH conditions where the two biopolymers are weakly electrically attracted to each other. In this study, the influence of polysaccharide linear charge density on the formation and properties of these biopolymer nanoparticles was examined. Mixed solutions of globular proteins (β-lactoglobulin) and anionic polysaccharides (high and low methoxyl pectin) were prepared. Micro-electrophoresis, dynamic light scattering, turbidity and atomic force microscopy (AFM) measurements were used to determine the influence of protein-to-polysaccharide mass ratio ($r$), solution pH, and heat treatment on biopolymer particle formation. Biopolymer nanoparticles ($d < 500$ nm) could be formed by heating protein-polysaccharide complexes at 83 °C for 15 min at pH 4.75 and $r = 2:1$ in the absence of added salt. The biopolymer particles formed were then subjected to pH and salt adjustment to determine their stability. The pH-stability of β-lactoglobulin-HMP complexes was greater than β-lactoglobulin-LMP complexes. The addition of 200 mM sodium chloride to heated complexes greatly improved the pH-stability of HMP complexes, but decreased the pH-stability of LMP complexes. The biopolymer particles formed consisted primarily of β-lactoglobulin, which was probably surrounded by a pectin coating at low pH values. AFM measurements indicated that the biopolymer nanoparticles formed were spheroid in shape. These biopolymer particles may be useful as delivery systems or fat mimetics.

**Fig. 6.4** Key elements of an abstract for a scientific manuscript, including context and objectives, materials and methods, results and discussion, and importance and relevance

novel plant protein that could be used as an ingredient to make plant-based foods, you might start with a general statement like "There is growing interest in replacing animal-based food ingredients with plant-based ones to reduce the adverse effects of the livestock industry on animal welfare, the environment, and human health."

- *Objectives*: The objective of the study should then be clearly and concisely stated so that the reader knows what particular topic or problem the manuscript focuses on. For the manuscript just mentioned, you might write "This study focuses on the isolation and characterization of proteins from duckweed, as well as their potential application as replacements for egg white proteins in plant-based egg substitutes."

- *Materials and methods*: A brief description of the key materials and methods used to carry out the research should then be given. Due to the limited space available, this description cannot include details about all the materials and methods used, but it should provide the reader with enough information that they know what was done and the main approaches used. For example, you might write "In this study, the potential of duckweed proteins to be used as egg substitutes was examined by characterizing their emulsifying and thermal gelation properties. Their emulsifying properties were characterized by examining their ability to form and stabilize oil-in-water emulsions, whereas their thermal gelation properties were characterized using dynamic shear rheology."

- *Results and discussion*: A concise description of the main manuscript findings should then be included, as well as a statement about the interpretation of the results. It is often useful to include quantitative data in an abstract. For instance, instead of saying that "small anionic oil droplets were formed when an aqueous solution of duckweed protein was homogenized with oil" you could say "small anionic oil droplets (200 nm, -25 mV) were formed when 90% w/w of an aqueous solution containing duckweed protein (1% w/w, pH 7.0) was homogenized with 10% w/w of corn oil using a microfluidizer (10,000 psi, 3 passes)." Then, if someone did not read your full paper, they could still use the data reported in your abstract in their own work.

- *Importance and implications*: Typically, it is useful to end an abstract with a sentence or two about the main findings, importance, and implications of the research. This might include a statement about how knowledge was advanced, how a particular problem was solved, or how the information obtained in the study could be used practically. It might also highlight areas where future work is still needed. For example, for the manuscript mentioned earlier, this might be "This study has shown that duckweed protein can be used as an effective emulsifier to form oil-in-water emulsions. Consequently, it may be used to replace animal-based emulsifiers in the food industry, thereby improving the sustainability of the food supply. Nevertheless, further research is needed to develop large-scale commercial processes to economically isolate duckweed protein in a pure form."

This structured approach is often useful for writing the first draft of an abstract. However, it may then need to be edited to make it clearer and more concise. For instance, it may be useful to group the materials, methods, results, and discussion used in one set of experiments together, and then talk about the same things for another set of experiments, rather than completely separating materials and methods from results and discussion. An example of an abstract written with the above structure is given in Fig. 6.4.

In summary, the abstract of the manuscript should address the following questions: What was done? Why was it done? What did you find? Why is it important? If a reader finds the answers to these questions relevant to their own work, then they may read your full manuscript. Here are some additional tips from the scientific writing website mentioned earlier (www.redwoodink.com):

- *Write the abstract last*: Usually, the abstract is written after all the other parts of the manuscript have been completed because then you have a better idea of the overall findings of the study.
- *Write clearly and concisely:* The abstract should be written as clearly, concisely, and compellingly as possible. Remove any words or phrases that are not essential. Avoid using acronyms, jargon, or technical terms that are difficult for readers to follow. Use active sentences where possible.
- *Focus on your own research*: Due to limited space, you should not discuss the work of others in detail. Instead, focus on the research you have carried out in your study. Citations are rarely included in abstracts.
- *Make sure the abstract follows journal guidelines*: Different journals have different formats and word limits for the abstract that authors must follow. Be sure that your abstract follows the guidelines of the journal you are submitting your manuscript to. Otherwise, it will be sent back to you, which will waste your and the editor's time, leading to a delay in the publication of your work.

- *Abbreviations*: When writing your abstract, it is important not to overuse abbreviations because this makes it difficult for readers to understand. If you do use abbreviations, then you should define them the first time you use them.

# Keywords

Many journals ask authors to include several keywords (typically 5–10) with their manuscripts to improve the visibility and impact of the work. Selecting appropriate keywords for your manuscript increases the chance that someone will find it when they are performing an online literature search. If you select impactful keywords, then more people are likely to find, read, and cite your manuscript, which will increase the impact of your research, as well as your reputation in the field. It is therefore important to choose your keywords carefully, rather than just using the first thing that comes into your head. Keywords should reflect the content of your manuscript and should be terms that are well known in your field of expertise. Some keywords are associated with "hot topic" areas, such as "nanotechnology", "sustainability", and "bioavailability" in my area (DJM). Consequently, it may be useful to include these in your list. However, if these terms already appear in the title or abstract of your manuscript, they may already be found by a search engine. For this reason, you may want to use synonyms of important words that you have already used to increase the chance of someone finding your manuscript during a literature search, like "nanoparticle", "environmental impact", and "absorption" instead of the three terms given earlier.

Some points to consider when selecting appropriate keywords are highlighted below:

- *Conform to journal instructions*: Make sure that the number and format of your keywords are consistent with the author guidelines of the journal you are submitting your manuscript to. Often, you can select your own keywords, but sometimes you must select them from a list specified by the journal.
- *Avoid overlap with title and abstract*: As mentioned earlier, a search engine may use keywords from the title and abstract, as well as from the keywords list. Consequently, you should use synonyms or alternatives to important terms or concepts that appeared in your title or abstract.
- *Don't use generic keywords*: It is important to use keywords that are not too general, or your paper may get lost among the enormous number of other publications in that area. For instance, if you were carrying out a study on the potential toxicity of titanium dioxide nanoparticles, you might use the keyword "inorganic nanoparticles" rather than just "nanoparticles" to help narrow the search.
- *Include information about materials and methods*: It is often useful to include some keywords about the main materials and methods used in your study. This helps researchers who are looking for studies that have been performed using particular substances or experimental methods. For instance, if your study used

differential scanning calorimetry to characterize the thermal properties of potato proteins, then you could include "differential scanning calorimetry" and "potato protein" in your keywords (or synonyms if you did not want to repeat words in the title or abstract).

# Introduction Section

The *Introduction* section of your manuscript should achieve several goals. It should provide readers with the background information needed to understand the study. It should state the reasons why the study was performed, what the specific problem addressed was, what your hypothesis was, what approach you intend to use to address the problem, and why the research is novel and important.

## *Structuring Your* Introduction *Section*

Writing the *Introduction* section is made easier by breaking it up into a series of key elements that are arranged in a specific order (Fig. 6.5):

### Context

You should usually start the *Introduction* section by providing the reader with some context about the purpose and nature of your study. Typically, an inverted pyramid approach is used, where you start by making more general statements and then continue by making more specific statements. Typically, you begin by stating the general area your research relates to. For instance, for a manuscript on the development of plant-based foods (like meat, seafood, egg, or milk alternatives), you might begin with a sentence like "There is increasing interest in developing plant-based foods because of concerns about the impact of animal-based foods on the environment, human health, and animal welfare. It has been reported that livestock animals are a major contributor to greenhouse gas production, pollution, and diversity loss, as well as to increases in certain kinds of chronic diseases." This information provides the reader with an understanding of the general problem that the research is designed to address.

After providing the big picture, you should narrow down the focus of your study by providing a statement that informs the reader about the specific topic of the manuscript. For instance, you might say something like: "The main focus of this article is to use soft matter physics principles to create plant-based meat analogs from proteins and polysaccharides derived from sustainable plant sources." This helps readers quickly establish what the manuscript is about and whether it is relevant to their work.

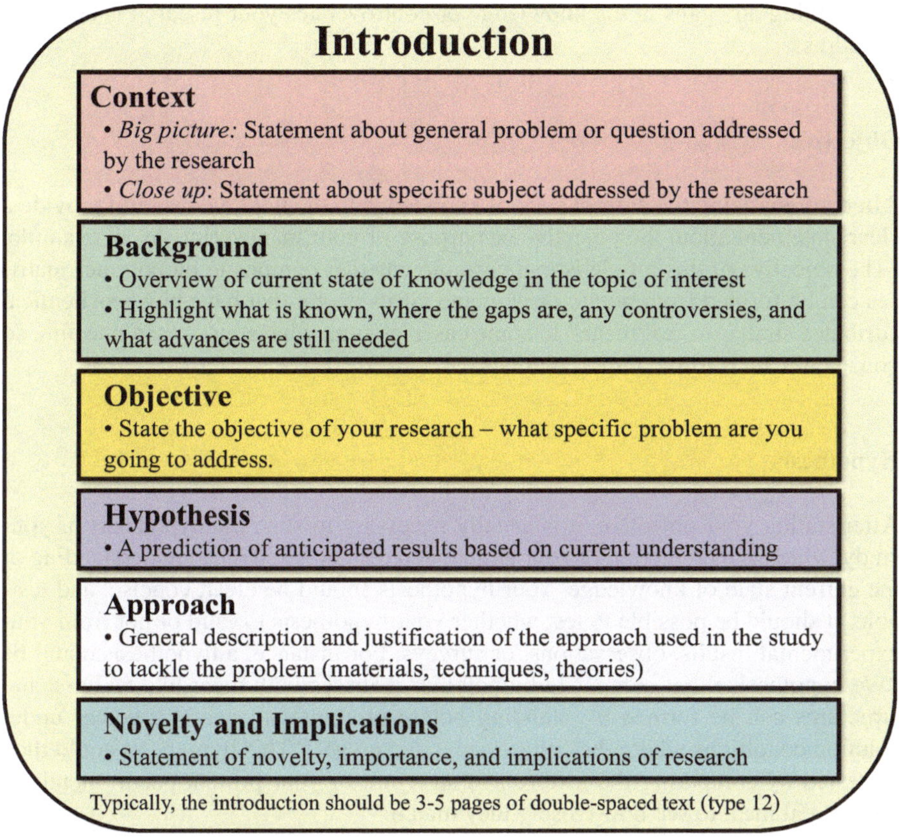

# Introduction

### Context
- *Big picture:* Statement about general problem or question addressed by the research
- *Close up*: Statement about specific subject addressed by the research

### Background
- Overview of current state of knowledge in the topic of interest
- Highlight what is known, where the gaps are, any controversies, and what advances are still needed

### Objective
- State the objective of your research – what specific problem are you going to address.

### Hypothesis
- A prediction of anticipated results based on current understanding

### Approach
- General description and justification of the approach used in the study to tackle the problem (materials, techniques, theories)

### Novelty and Implications
- Statement of novelty, importance, and implications of research

Typically, the introduction should be 3-5 pages of double-spaced text (type 12)

**Fig. 6.5** Key elements of the *Introduction* section of a scientific manuscript, including context, background, objective, hypothesis, approach, novelty, and implications

## Background

Once you have narrowed the focus of your manuscript onto a particular subject, you should then provide a brief overview of the current state of knowledge in that area. Because of the limited space available for the *Introduction* section (typically 3–5 pages), you cannot give a comprehensive and detailed review of all literature that has ever been published on the topic. Instead, you can refer to the most relevant original research articles to your study, as well as to original research articles that are closely related to your study. It is important to give a balanced overview of the relevant background literature – you should not just cite manuscripts you or your research group has published. Instead, you should try to cite all the major players in the field, as these people may review your manuscript. If you ignore their work, they may be upset and more likely to find reasons to reject your paper. In summary, this section should provide an overview of what is currently known in the area, highlighting any gaps in the knowledge or controversies your research is designed to address.

## Objective

After summarizing the current state of knowledge in the field, you should provide a clear statement about the objective or purpose of your manuscript. As an example, "The objective of this article is to determine whether composite biopolymer matrices can be formed from potato protein and citrus pectin that have physicochemical attributes similar to real meat." In some cases, you may have more than one aim, so you should state all the important ones.

## Hypothesis

After stating your objective, it is usually necessary to state the hypothesis of your study, which is a prediction of your anticipated results based on an understanding of the current state of knowledge. Your hypothesis should be clear, concise, and testable. It should be possible to test whether your hypothesis is valid or not from your experimental results, observations, or surveys. For instance, a hypothesis could be "We hypothesize that composite biopolymer matrices with meat-like textures and structures can be formed by blending potato proteins and pectin together under solution conditions where they phase separate and gel." This hypothesis could then be tested by comparing the texture and structure of your protein-pectin blends to that of real meat to see how closely they match.

## Approach

After stating your hypothesis, you should give a brief overview of the approach you intend to use to tackle the problem. At this point, it is not necessary to go into detail about the materials and methods used, but you should provide a brief overview of their key characteristics and your rationale for using them. For example, "Potato proteins are water-soluble globular proteins that are typically isolated from the wastewater produced by the potato industry (Smith et al., 2022). This wastewater contains two main protein fractions: patatin, which is a glycoprotein (39–43 kDa), and a protease inhibitor fraction, which is a mixture of globular proteins (5–35 kDa). In this study, patatin was used as a model potato protein because it forms strong gels when heated above its thermal denaturation temperature." Suitable citations should be given to these statements where needed. This brief overview of key materials and methods should be designed so that the reader can understand the approach used and why it was adopted.

**Novelty and Implications**

Toward the end of your *Introduction* section, you should highlight the novelty and implications of your research study. How is it different from other people's work? How might it advance the understanding of the field? What gaps in knowledge is it intended to fill? How will it benefit society? As an example, you might end your *Introduction* section with a statement like: "To the author's knowledge, this is the first manuscript to use soft matter physics principles to create plant-based meat substitutes using potato protein and citrus polysaccharide composites. This information may be useful for creating healthier and more sustainable plant-based foods, thereby reducing the negative impacts of the modern food supply on human health and the environment." However, it is important that you have performed a thorough search of the literature to ensure that this kind of statement is correct.

## *Citing Previous Research*

When deciding which articles to cite in your *Introduction section*, you should consider the following:

- *Balanced*: Ensure you have cited the most important and relevant research in the field, including all the people and research groups who have made important contributions. If an area is controversial, make sure you cite articles on both sides of the argument.
- *Current*: Ideally, most of your citations should be from research published in the past decade or so. However, sometimes, it may be important to cite older articles if they were the first to report an important discovery or if the theory or technique reported in those articles is still commonly used in the field.
- *Relevant*: It is important to only cite manuscripts that are relevant to your study. Do not discuss research that is only tangential to your own study, as this may confuse the reader and waste space.

## Materials and Methods Section

The purpose of the *Materials and Methods* section is to provide the reader with detailed information about how the study was carried out. Typically, this section is divided into several subsections. The first subsection provides information about all the reagents and other substances used. The intermediate subsections describe the various experimental protocols used. The final subsection contains information about the statistical methods used for data analysis.

## *Structuring Your* **Materials and Methods** *Section*

The various subsections in the *Materials and Methods* section are described in more detail below and shown schematically in Fig. 6.6.

### Materials Subsection

The first subsection should provide clear information about all the materials used in the study, including chemical reagents, buffers, solvents, polymers, enzymes, cells, cell culture media, microscopy stains, water, and test samples. For each material, it is important to provide a clear statement about precisely what it is and where it was obtained so that other people could use the same material in their study. For instance, one might state "Sodium chloride (S9888; >99.0% pure) was obtained from the Sigma Chemical Company (St. Louis, MO, USA)." The name, city, state, and country of the company or institution that provided the materials should be included. Water is commonly used in scientific studies as a solvent or reagent. Water can vary greatly in purity and properties depending on where it came from and how it was treated. Consequently, it is important to state the precise nature of the water used in your study, such as "Double distilled water obtained from a water purification system was used in this study, which had an electrical resistance of 18 MΩ·cm." If you

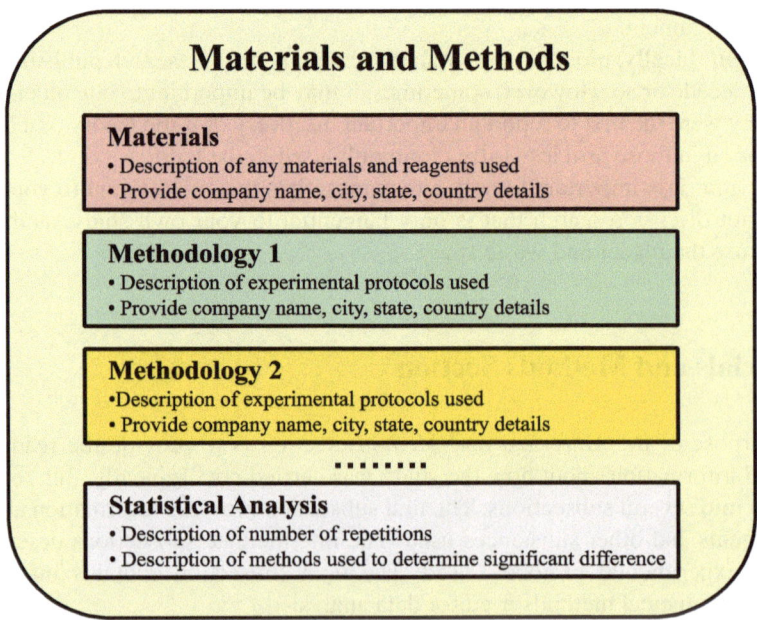

**Fig. 6.6** The *Materials and Methods* section of a scientific manuscript typically includes *Materials*, *Methodologies*, and *Statistical Analysis* subsections

are using clinical samples, such as human blood, urine, or feces, then you may have to include a statement about them being collected under approved ethical guidelines.

## Methodology Subsections

Several *Methodology* subsections typically follow the *Materials* subsection, which provide clear and precise instructions on the different approaches used to perform each part of the study. Each subsection should describe the protocols and equipment used to prepare, test, and analyze the samples, and be ordered in a sequence that follows the timeline the experiments were performed (like a recipe). Appropriate citations should be given if a method was adopted from a previous study, such as "The method used to measure the protein content of the samples was based on one described previously (McClements et al., 2023)." The instrument model, company name, city, state, and country should be included for each piece of major equipment used to prepare or characterize the samples so that other people know what equipment was used and who they can purchase it from. As an example, "The particle size distribution of the emulsions was measured using a laser diffraction instrument (MasterSizer 3000, Malvern Instruments, Malvern, Worcestershire, UK)." In each *Methodology* subsection, it is also important to provide information about the number of repetitions carried out. For instance, "Each sample was prepared and tested three times, with two repeat measurements made per sample, leading to a total of six data points." If you are using aqueous solutions in your studies, you should report the pH and ionic composition. For example, "A protein solution was prepared by mixing 10.0 g of potato protein in a buffer solution (10 mM phosphate, pH 7.0)." In some cases, the *Methodology* subsections may follow the same order as the subjections described in the *Results and Discussion* section to increase clarity.

## Statistical Analysis Subsection

The final subsection should provide information about the statistical methods used for data analysis. If the information was not provided in earlier sections, then this section should state the number of repeated measurements that were carried out to calculate any means and standard deviations. It should also provide information about any methods used to establish statistically significant differences among samples and the level of confidence used. For instance, "The properties of each sample were measured in three independent trials and the mean and standard deviation of these measurements were calculated. ANOVA (post hoc Tukey HSD test) was used to establish significant differences between samples at a 95% confidence level ($p < 0.05$) using a statistical software program (SAS/STAT, Cary, North Carolina)".

## *Other Tips*

Typically, the *Materials and Methods* section is written in the past tense. If you are using newly developed methods, then it is important to provide enough detail so that someone can repeat your work. However, if the methods are already established and have been published in detail elsewhere, then you only need to provide a brief description and cite the previous study. As an example, one could state "The method used to characterize the size of the particles in the samples has been described in detail previously (McClements et al., 2023)."

Commonly, the *Materials and Methods* section goes immediately after the *Introduction* section of a manuscript, but this is not always the case. Some journals include it at the end of the manuscript so that readers can get to the *Results and Discussion* section earlier (which is usually the most important part). It is therefore important to check the "Instructions for Authors" or "Author Guidelines" of the journal where you intend to submit your manuscript to ensure you are using the correct format. Alternatively, you can look at other manuscripts that have been published in your target journal to see what format they used when writing their *Materials and Methods* section.

## Results and Discussion Section

The *Results and Discussion* section of a manuscript is usually the most important part because it contains any new findings, as well as their interpretation. Some of the most important factors to consider when writing the *Results and Discussion* section are therefore considered here.

## *Objectives of the Results and Discussion Section*

There are several objectives to keep in mind when writing the *Results and Discussion* section of a scientific manuscript:

- To clearly and precisely present, describe, and analyze the data obtained in the study.
- To provide a rational interpretation of the data that is supported by experimental evidence. This may involve interpreting the data in terms of known physical, chemical, biological, or psychological principles, fitting the data to theoretical models, or modeling the data using computer simulations.
- To show how the new findings are related to previous knowledge and how they advance the field of study.

## *Using the Correct Format*

Some academic journals require a combined *Results and Discussion* section, others require a separate *Results* section and *Discussion* section, and others give the author the flexibility to decide. It is therefore important to understand the required format for the journal you intend to submit your manuscript to. You can find this by reading the "Instructions to Authors" provided by the journal or by looking at the structure of other manuscripts published in the target journal. When there is a choice, we prefer combining the *Results* and *Discussion* sections because we feel it improves the clarity of the writing and avoids repetition, but other people prefer to have them separately. One argument for having the two sections separate is that the results are considered "hard data" that can be interpreted in different ways. Consequently, a reader can see the results and make their own judgments without being influenced by any interpretations made by the author.

## *Structuring Your* Results and Discussion *Section*

Usually, the *Results and Discussion* section should be split into several subsections, where each one deals with a particular group of experiments or observations. The number of subsections varies depending on the nature of the study and the writing preferences of the authors. Typically, 3–5 major subsections are a reasonable number, which may contain their own subsubsections. The most important factor to consider when deciding on the nature of the subsections used is the clarity of the writing. You are trying to organize your subsections to make the manuscript as clear, concise, and compelling as possible. In cases where the subsections in the *Results and Discussion* section are similar to those in the *Materials and Methods* section, it is good practice to follow the same order in both (Fig. 6.7), as this allows the reader to easily find the appropriate protocols for each experiment. However, this is less important when subsections in the *Results and Discussion* section are different from those in the *Materials and Methods* section. In some cases, it is advisable to order your subsections so that the findings are discussed in order from most to least importance, but this depends on the subject matter.

There are several approaches to deciding what topics are covered in each subsection, with some of the most common given here:

- *Themes*: The subsections can be organized into different general themes related to groups of experiments performed in the study. For instance, for a study on the development of nanoparticle-based delivery systems for bioactive agents, the subsections could be "Fabrication and characterization of nanoparticles," "Impact of environmental conditions on nanoparticle stability," and "Simulated gastrointestinal fate of nanoparticles." This approach is probably the most commonly employed, as it can improve the organization and clarity of manuscripts.

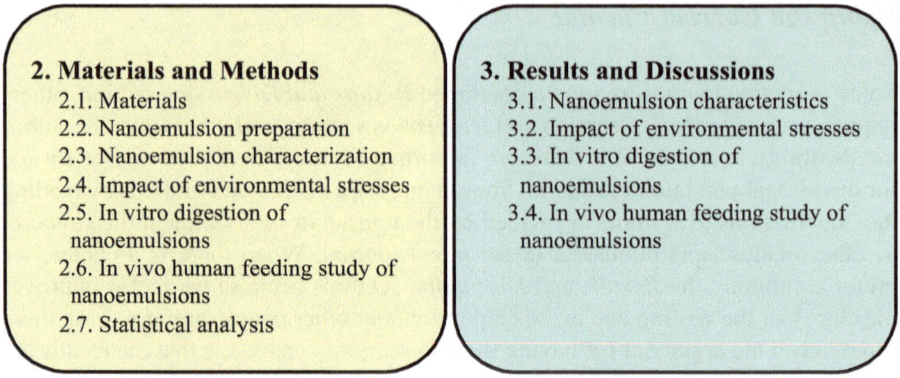

**2. Materials and Methods**
  2.1. Materials
  2.2. Nanoemulsion preparation
  2.3. Nanoemulsion characterization
  2.4. Impact of environmental stresses
  2.5. In vitro digestion of
       nanoemulsions
  2.6. In vivo human feeding study of
       nanoemulsions
  2.7. Statistical analysis

**3. Results and Discussions**
  3.1. Nanoemulsion characteristics
  3.2. Impact of environmental stresses
  3.3. In vitro digestion of
       nanoemulsions
  3.4. In vivo human feeding study of
       nanoemulsions

**Fig. 6.7** In some manuscripts, it is advisable to have corresponding subsections in the *Results and Discussion* section and the *Materials and Methods* section to improve the clarity of your manuscript

- *Methods*: The subsections are sometimes organized according to the major methods used to obtain the data. For instance, this could be "Particle size and charge characteristics," "Particle morphology analysis," "Fast Fourier transform analysis of molecular interactions," and "X-ray diffraction analysis of particle crystallinity." This approach is useful when a researcher is preparing a particular material (or group of materials) and then using a variety of analytical tools to characterize it (them).
- *Combined*: Sometimes it is useful to organize the subsections into general themes, and then in each theme have subsubsections that discuss the data obtained using different analytical methods. This combines the advantages of the two approaches.

The most important thing when designing your *Results and Discussion* section is to have subsections that tell a clear story that engages the reader and presents the data in an accurate and concise manner.

## Subsection Structure

Writing the *Results and Discussion* section can be facilitated by writing each subsection in a standardized format (Fig. 6.8).

- *Purpose statement*: Typically, it is advantageous to start each subsection with a statement about the purpose of the experiments to orientate the reader. For instance, "In this section, the purpose of the experiments was to determine the impact of solution conditions on the aggregation stability of protein nanoparticles."
- *Approach statement*: It is then useful to state the general approach that was taken. For example, "The impact of salt, pH, and temperature on nanoparticle aggrega-

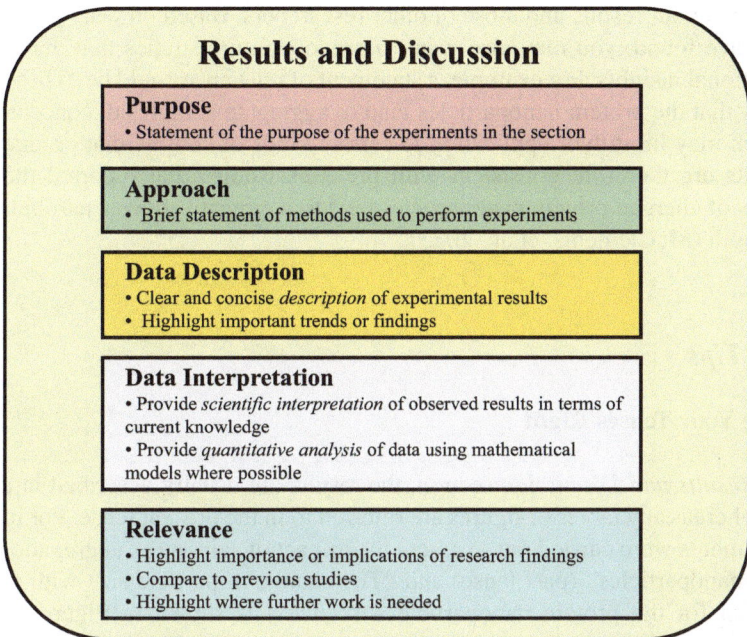

**Fig. 6.8** Writing the *Results and Discussion* section can be facilitated by using a standardized format to write each subsection

tion was characterized by measuring changes in particle size using a dynamic light scattering instrument."

- *Data description*: The data obtained from the experiments should then be described and any trends observed in the results should be highlighted. For instance, "The dynamic light scattering measurements showed that the size of the particles in the protein nanoparticle suspension increased as the salt concentration increased."
- *Data interpretation*: After all the relevant data have been presented, usually in the form of graphs and tables, it is necessary to interpret the data using appropriate scientific principles, mathematical theories, or computer simulations. For example, "The observed increase in particle size with increasing salt concentration can be attributed to the aggregation of the protein nanoparticles caused by electrostatic screening effects, *i.e.,* the accumulation of $Na^+$ ions around the anionic groups on the protein surfaces."
- *Relevance*: At the end of each section, a statement should be included about the potential relevance of the findings. Are your findings consistent with those of previous studies? Do they resolve current controversies in the literature? Do they lead to a new understanding of the phenomenon investigated? This typically involves comparing your data with that reported by other researchers on closely related systems, and critically stating and discussing any similarities or differences. It is useful to give potential reasons for any discrepancies observed

between your results and those of other researchers. Based on what you and others have found, you may also want to propose future studies that may provide additional insights. For example, a statement of relevance could be "These results show that the protein nanoparticles tend to aggregate at high salt concentrations, which may limit their application in certain kinds of commercial products. Our results are therefore consistent with previous studies that reported that other kinds of charged protein nanoparticles tend to aggregate above a particular ionic strength (McClements, et al., 2023)."

## *Other Tips*

### Getting Your Tenses Right

In the *Results and Discussion* section, the results are usually described in the past tense, whereas any tables or figures are referred to in the present tense. For instance, "Experiments were carried out to assess the impact of salt on the aggregation of the protein nanoparticles" (past tense) and "The change in particle size with salt concentration for the protein nanoparticle suspension is shown in Figure 1" (present tense).

### Avoid Duplication

It is important not to duplicate the same data in the tables, figures, and text. If you have already summarized the data in a table, then you don't need to provide the numbers again in the text. You can simply refer to the table and then provide a discussion of the general trends observed in the text. Most journals don't allow you to present the same data as a table and as a figure, you must choose one or the other. Tables are good because they allow other researchers to have access to all the numerical values (*e.g.,* 1432.3 ± 1.3 m/s), so they can use the data in their own work (provided they cite your work properly). However, figures are often useful because they show trends in the data that can easily be discerned visually. In some journals, it is now possible to include supplementary information. This is information that does not appear in the main manuscript but can be accessed by readers online. In this case, it may be possible to show the data as a graph in the main article to increase its visual impact, and then include the data as a table in the supplementary information so that other people have access to it. If you do not have a lot of data from an experiment (*e.g.,* just two or three data points), then it's better to include them as numbers in the text rather than use a figure or table. Tips on preparing figures and tables are given in Chap. 5.

## Being Statistical

When you are comparing the measurements made on different samples, it is important to provide information about whether any differences observed are statistically significant (Chap. 5). In this case, you can include information about the level of confidence; for instance, for 95% confidence $p$ is less than 0.05. As an example, "the mean diameter of the particles measured at 200 mM NaCl (2230 ± 23 nm) was significantly higher than that measured at 50 mM NaCl (882.3 ± 0.9 nm) ($p < 0.05$)."

## Beware of Overinterpretation and Overstatement

Ideally, you should give a detailed interpretation of your experimental findings in terms of basic scientific principles. It is important, however, not to overinterpret your results. In particular, you should not make speculations that are not supported by your experimental data. It is also important that you do not overstate your findings, *e.g.*, by saying something like "This conclusively proves ...", when you do not have sufficient evidence to make that claim.

## Don't Waffle

You should present and interpret your results in a clear, concise, and unambiguous fashion. Don't give long-winded and circuitous descriptions or interpretations of your results, as this will bore and confuse readers.

## Highlight Limitations

It is often useful to highlight the limitations of your study as this shows reviewers and readers that you have considered potential weaknesses in the design and interpretation of your experiments. No study can be completely comprehensive, and it is always necessary to make some compromises in the experimental design. Highlighting these limitations lets the reviewers and readers know that you understand the big picture. It may also help to point out where additional experiments are needed in the future.

# Conclusions Section

The last main section of a scientific manuscript is the *Conclusions* section. This section should briefly summarize the main findings of the manuscript, as well as highlighting their broader significance. Again, it is useful to adopt a consistent structure

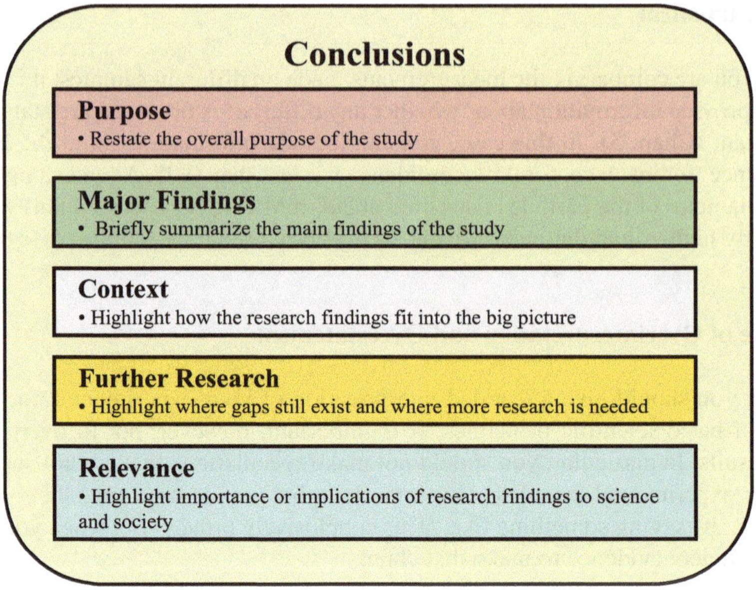

**Fig. 6.9** Writing the *Conclusions* section can be facilitated by dividing it into several elements, such as purpose, major findings, further research, context, and relevance

when writing the *Conclusions* section as this makes it easier to write and read (Fig. 6.9).

## *Purpose*

Initially, you should briefly restate the purpose or hypothesis of the study to remind the reader of the problem that was addressed. For instance, "The purpose of this study was to determine whether antisolvent precipitation could be used to fabricate protein nanoparticles" or "This study was designed to test the hypothesis that anti-solvent precipitation could be used to fabricate stable protein nanoparticles."

## *Major Findings*

The major findings of the study should then be summarized. In addition, you should state whether your initial hypothesis was supported by your findings or not. It is not necessary to provide a detailed discussion of your findings, as this has already been done in the *Results and Discussion* section, and partly in the *Abstract*. Instead, you should just give a general overview of the most important findings of the study. For instance, "It was shown that pea protein nanoparticles could be successfully fabricated using a simple antisolvent precipitation method, provided that the preparation conditions were optimized."

## *Context*

Next, you should highlight how your findings fit into the big picture. For example, "The protein nanoparticles developed in this study may be viable alternatives to existing nanoparticles used for the encapsulation, protection, and delivery of hydrophobic bioactive substances, such as nanoemulsions or nanoliposomes."

## *Further Research*

Then, you should point out any limitations of the study and where further research is still needed. For instance, "In the future, it will be important to demonstrate that this method can be adopted for large-scale economic production of protein nanoparticles using food-grade ingredients," or "In the future, it will be important to test the efficacy and safety of these protein nanoparticles using *in vivo* animal feeding studies."

## *Relevance*

Finally, you should highlight the importance of your findings and how they advance the understanding of the area. For instance, "We have shown that protein nanoparticles can be produced using a simple and inexpensive fabrication method, which may provide a useful alternative to existing methods." You should also highlight the importance of your findings to society as a whole. For example, "The development of protein nanoparticles that can be used to encapsulate, protect, and deliver bioactive agents may lead to the creation of a new generation of functional foods and medicines that can promote human health and well-being."

In summary, the *Conclusions section* should be concise and provide reviewers and readers with a clear take-home message. For example, why the study was important, what the main findings were, what still needs to be done, and why the findings are important to the scientific community and wider society.

## Acknowledgments Section

In the *Acknowledgments* section, you should thank anyone who has helped you carry out the study, but whose contribution is not sufficient to be included as a coauthor. These could be people who trained you to use a particular analytical procedure, people who you discussed your study design with, or people who have read your paper and provided constructive comments or suggestions. You should also acknowledge any institutions that provided financial support for your research, such as government agencies, nonprofit organizations, or industrial sponsors. In many cases, you may have to include a specific code for the grant you obtained to fund the study. You may also thank companies that supplied you with materials or instruments used in your study.

## Declaration of Competing Interests

Most scientific journals require authors to include a "Declaration of Competing Interests" statement to disclose any potential conflicts of interest that could affect the validity of the study. There are numerous potential competing interests that might fall into this category, which may be financial or nonfinancial conflicts of interest:

- Being paid as a consultant for a company that might benefit from the research.
- Receiving funding from a company whose products are related to the research.
- Serving on a scientific advisory board for a company whose products are related to the research.
- Working for an advocacy group.

If you and your coauthors have no known competing interests, you can simply state "The authors declare no conflicts of interest." Conversely, if one or more of the authors do have competing interests you could state "David Julian McClements is a paid consultant for company X, which manufactures plant-based chicken products."

## References Section

The final part of the manuscript is the *Reference* section, which includes a list of all the sources of information you cited, such as books, book chapters, scientific manuscripts, patents, abstracts, magazine articles, theses, online databases, *etc*. For each scientific manuscript you cite, you typically need to provide information about the authors, publication year, article title, journal title, volume, issue, and page numbers. For books, you usually provide information about the authors, publication year, book title, publisher, and publication place. However, the precise format in which this information is presented varies from journal to journal. Consequently, it is important to carefully read the "Instructions for Authors" of your target journal so that you use the correct reference format. If you use an incorrect format, the journal may return the manuscript to you without reviewing it, which will delay the publication process. As mentioned in an earlier chapter, the reference list should include all the key articles in the field of study with a good balance of the key researchers. This is important because the people who review your article may also work in the field, and they may be upset if you have not referenced their work. As a result, they may be more critical of your manuscript when they review it. Ideally, reviewers should assess manuscripts objectively without taking their personal feelings into account, but this ideal is often difficult to achieve in practice. You should also be careful not to cite too many papers from your own lab group, as reviewers may object to this also. Indeed, some scientific journals now have a limit to the percentage of papers that can be cited from the authors' own group. If you are citing authors by name and year and they have published more than one article in a certain year, then you may have to distinguish between the articles by using letters, *e.g.,* McClements, 2023a, 20023b or McClements 2023a,b. The preparation of the reference section can be facilitated by using specialized reference manager software, thereby greatly enhancing your research productivity.

## *Reference Managers*

Reference managers are software programs that help you store, organize, and cite your references. They are incredibly powerful tools for writing scientific manuscripts because they save considerable time in finding and citing other studies. You should therefore find a suitable reference manager at the beginning of your graduate studies. There are numerous reference managers available, including EndNote, Mendeley, RefWorks, and Zotero. Some of these programs are free, whereas others must be purchased (students often receive a discount on the full price or universities may have a license to use them). From our experience, the benefits of using these programs far outweigh the costs. One of the most important things you should consider when selecting a reference manager is which one your professor and other members of your laboratory use, as well as any people you may be collaborating with. Ideally, everyone in your research group should use the same reference manager as this will greatly facilitate the writing and editing of manuscripts and graduate theses.

In this section, we use EndNote as an example to demonstrate the advantages of using a reference manager. Typically, you should start a new EndNote file for each manuscript you are writing and give it an informative name related to the subject matter of the manuscript, such as "PB Foods" for a paper on plant-based foods. In this file, you should store all the references that are relevant to your study and will be cited in your final manuscript. The next step is to locate and download the references you intend to use. You typically do this by using a literature search engine like

PubMed, Google Scholar, or Web of Science. After entering the appropriate keywords or author names, the search engine generates the titles of relevant scientific manuscripts and other documents. After reading the titles and abstracts of these manuscripts, you select the ones you want to download by checking the checkbox in the program. All the relevant citation information from the manuscripts you have selected, such as the title, year, authors, journal, issue, page numbers, and abstract, is then downloaded to your computer into the EndNote file you currently have open. Every time you find a new relevant article, you can download it to the same file. Thus, the first advantage of using these software programs is that they help you keep all the relevant manuscripts together in one place. The second advantage is that they allow you to simply add citations into the manuscript you are writing. Typically, you select one or more of the manuscripts you want to cite in the EndNote file and then go into your word processing file.[3] You can then easily cite the manuscripts by clicking on the appropriate button in the word processing program (Fig. 6.10). Once you

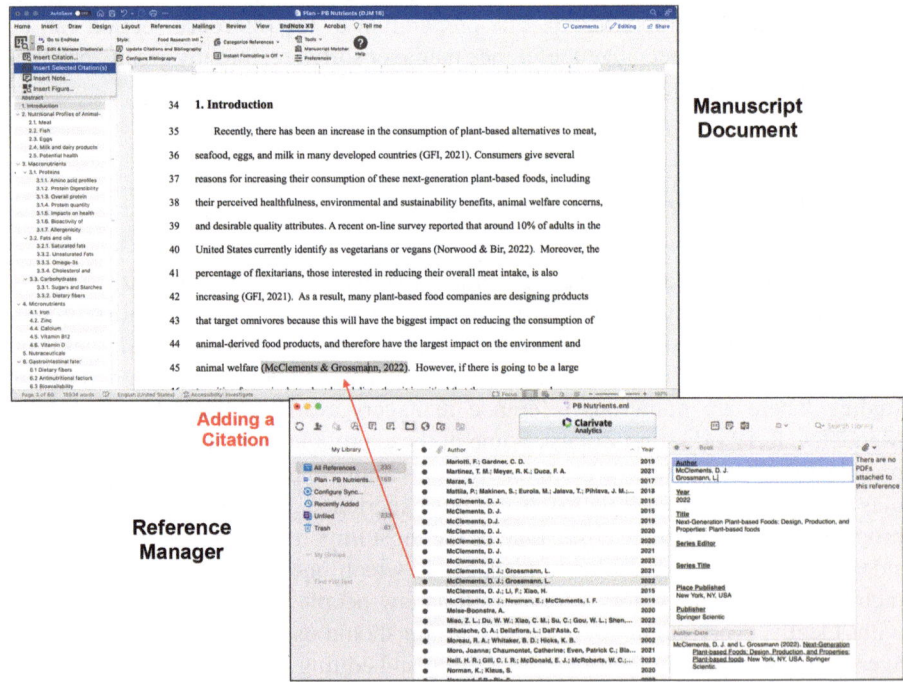

**Fig. 6.10** Reference managers are extremely convenient tools for collecting, storing, and citing references for scientific manuscripts or theses. Here, a citation is being added from a reference manager (EndNote) file to a manuscript (Word) file

[3] Typically, you have to associate the reference manager with the word processing document first. For instance, in Microsoft Word you would add ENDNOTE to the Templates and Add-ons in the Tools menu.

have cited the articles in your manuscript, you can format them using preexisting templates by simply clicking another button. This is extremely convenient because each scientific journal has its own reference formatting style. Thus, if you are not sure where you will submit your manuscript, or if your manuscript gets rejected from one journal and you want to submit it to another one, you can easily reformat the references with a simple click of a button. This can save considerable time and energy.

To summarize, there are several steps involved:

- *Create reference file*: Create a new reference file when you start writing a new document, such as a manuscript, chapter, or thesis.
- *Locate and download references*: Locate and download the citation information of the articles you want to cite in your document using a suitable literature search engine (such as Web of Science).
- *Cite references*: Cite the references in your document.
- *Format references*: Chose an appropriate format based on the Author Guidelines for your document, *e.g.,* a specific scientific journal.

Each reference manager has its own particular format, but they all have similar general features that allow you to store, cite, and format references.

# Writing Review Articles

Review articles provide an overview and synthesis of existing research on a specific topic. As discussed in an earlier chapter, it is beneficial to write a review article during the early stages of your graduate studies. This helps you to develop a thorough understanding of your research area and increase your visibility in the field. Review articles are usually used and cited more than original research articles, thereby increasing the impact of your work. Review articles have a different objective and are written in a different style than original research articles. For this reason, we provide some additional tips on how to prepare review articles in this section.

- *Select an appropriate topic*: Choose a topic that is relevant, interesting, and has enough published research to support a comprehensive review. Typically, there should be more than approximately 50 original research articles published in the field before a review article can be written. The topic should also be directly related to the interests of your research topic, as the information collected may also be used to write the introduction to your thesis. Before starting to write, you should carry out a preliminary review of the literature to be sure no one else has written a review article on a similar topic recently. If they have, you may need to change the emphasis of your article.
- *Plan and organize*: Before you begin writing, create a clear outline that includes the main sections and subsections of your review article. This will help you structure your thoughts and ensure a logical flow of information. Typically, the

main sections in a scientific review article are the abstract, introduction, main body, conclusion, and references. However, other sections can also be included, such as a methodology section that describes the methods used to identify and collect the articles included in the review. For the main body section, it is a good idea to plan the different subsection topics before carrying out an extensive literature review (but these can be revised as you acquire more information).

- *Conduct a thorough literature review*: Familiarize yourself with the existing research on the topic. Read primary research articles, review articles, and other relevant sources. Take detailed notes and identify key findings, methodologies, and gaps in knowledge.
- *Write a compelling introduction*: The *Introduction* section of a review article should follow a fairly similar pattern to that of an original research article (Fig. 6.5). It should start by stating the topic that is to be addressed (context), and then provide an overview of the current state of the knowledge in the field (context). It should then state the purpose and scope of the review (objective and approach). Finally, it should highlight why the review article is different from previous studies (novelty), as well as the importance of the information presented (implications).
- *Organize the main body*: Divide the main body into logical sections or themes based on the subtopics or aspects you want to cover. Present the information in a systematic and coherent manner using clear headings and subheadings to guide the reader.
- *Synthesizing and analyzing the literature*: Rather than simply summarizing individual studies, strive to synthesize the findings from multiple sources. Identify common trends, patterns, controversies, and areas of agreement or disagreement in the literature. Critically analyze the strengths and weaknesses of the studies reviewed.
- *Be objective and unbiased*: Maintain objectivity throughout your review article. Present different viewpoints, conflicting evidence, and limitations of the studies reviewed. Avoid personal bias and provide a balanced perspective.
- *Provide clear conclusions*: Summarize the main findings of your review article and relate them to your research questions and objectives. Identify the implications of the findings and suggest areas for future research.
- *Use proper referencing and citation style*: Provide a balance of citations from all the major research groups working in the field. Don't overly rely on citations from your own research group. Accurately cite all the sources you have used. Follow the specific referencing style needed by the target journal or publication guidelines. Using a reference manager will greatly facilitate finding, storing, and citing references.
- *Use informative figures and diagrams*: Prepare figures or diagrams that highlight the most important aspects of the material covered in the review – pictures are usually much more impactful and memorable than words. You may use figures or diagrams from other people's studies, provided you have obtained the copyright permissions from the publisher. This can often be done online from the publisher's website.

- *Revise and edit*: Review your article for clarity, coherence, and flow. Ensure that your arguments are well supported, and the content is free of grammatical and typographical errors. Seek feedback from colleagues or mentors before submitting your manuscript.

Writing a good scientific review article requires a balance between presenting existing knowledge and adding your unique perspective. As with original research articles, you want to tell a compelling story that will interest potential readers. It should be noted that some journals require an invitation before a review article can be submitted, and so you should check the Author Guidelines before submitting your manuscript.

# Writing Tips

In this section, we provide a few general tips for improving the overall quality of your scientific writing.

## *Understand Your Audience*

You should consider who will be reading your manuscript and tailor your writing style and language to suit their level of expertise. Ensure that your writing is accessible and understandable to the intended audience.

## *Develop a Strong Narrative*

Randy Olson has written several books for scientists that encourage them to improve their communication skills. There is no point doing great science if no one knows about it. He is in a unique position to advise scientists on communication. After doing his graduate studies at Harvard University, he became a Professor of Marine Biology at the University of Maine. He was granted tenure (a guaranteed job for life!) but gave it up to become a film director in Hollywood. Since then, he has made several highly acclaimed films and written numerous successful books. In "Houston, We Have a Narrative" (2015), he highlights the critical role that narrative plays in scientific writing. Most scientists are not taught to be good storytellers, which means that scientific manuscripts are often dull and/or confusing, thereby reducing their impact. When writing a scientific manuscript, you should aim to tell a compelling story, especially in the introduction, discussion, and conclusion sections. Why is the research topic you are working on important? What did you do? How did you do it? Why did you do it? What did you find? Why did this happen? Why are your results important? What are the implications of your findings? A good scientific manuscript should answer all these questions. It should also be written in a way that is clear, concise, and engages the reader. Olson and his coworkers have developed a simple strategy for creating more effective and compelling scientific writing, which they call ABT ("And" "But" "Therefore"), which is summarized in Chap. 9. Briefly, this method makes a story more compelling by providing the current context ("And"), highlighting a problem that exists ("But"), and then describing how the problem is resolved ("Then"). The details of the ABT method can be found in a recent book called The Narrative Gym: For Science Graduate Students and Postdocs (2022) by Douglas, Bahr, and Olson (see Resources). We would certainly recommend reading this short book as it provides some excellent advice on becoming a more effective scientific writer.

## Be Clear and Precise

Write in a straightforward and concise manner. Typically, space is a premium in scientific journals, so it is vital to tell your story using as few words as possible while still being clear, precise, and compelling.

- *Clear*: The structure of your sentences should make their meaning absolutely clear – avoid any potential ambiguity by constructing your sentences well. Avoid everyday speech, jargon, and excessive use of acronyms as this can make your manuscript difficult to read and understand.
- *Precise*: Use as few words and sentences as possible to describe and discuss your work, while still getting your meaning across.

We give an example here that shows how sentences can often be reduced substantially by eliminating unnecessary words or phrases without losing any meaning:

- "We carried out a literature search and were excited to find a paper that reported that the bioavailability of $\beta$-carotene depends on the type and amount of fat present (Qian et al., 2014)." [29 words]
- "The bioavailability of $\beta$-carotene has been reported to depend on the type and amount of fat present (Qian et al., 2014). [17 words]
- "$\beta$-carotene bioavailability depends on fat type and amount (Qian et al., 2014). [8 words]

The art of good scientific writing is to be as clear and precise as possible, without losing the intended meaning.

## *Spelling and Grammar*

Your scientific manuscript should be free of spelling mistakes and grammatical errors. If you have a lot of these in your manuscript, then reviewers may think you are careless and therefore have less confidence in your research findings. It is much easier to make spelling mistakes and grammatical errors (and more difficult to spot them) if the language you are writing in is not your native tongue. In most cases, this occurs when English is a second language. Several strategies can be used to detect these errors. You can have a native speaker read through and correct any errors, provided you can find someone who has the time and willingness to do this. The most common word processing programs (such as Microsoft Word) have built-in spelling and grammar checkers that can highlight mistakes in the text and suggest corrections. It is therefore important to ensure that these tools are switched on when you are writing and editing your manuscripts. There are also specialized programs that can be used to check for spelling and grammatical mistakes, such as Grammarly, which usually require a subscription fee for premium access but can be valuable for improving the quality of your writing. Additionally, some artificial intelligence programs, such as ChatGPT, can read over manuscripts and identify spelling or grammar mistakes.

## *Be Consistent in Your Writing Style*

You should maintain a consistent writing style throughout your manuscript. Typically, it is best to use an active voice, the present tense, and a formal tone. You should avoid excessive use of the passive voice or convoluted sentence structures.

*Voice*: The passive voice is often used in scientific writing to emphasize the object or process being studied rather than the person or entity performing the action. It is generally the most appropriate form to use in the *Materials and Methods* section. For example, "The samples were analyzed using high-performance liquid chromatography (HPLC)." However, it is typically recommended to use the active voice whenever possible to improve the clarity and readability of your writing; particularly, in the *Introduction, Results and Discussion*, and *Conclusions* sections. For instance, you may say "We found a statistically significant increase in the lifespan of fruit flies treated with the nutraceutical compared to the control group" (active) rather than "A statistically significant increase in the lifespan of fruit flies treated with the nutraceutical compared to the control group was found." The active voice is often preferred in scientific writing because it tends to be more direct, concise, and easier to understand.

## Seek Feedback

Before submitting your work, seek feedback from colleagues or mentors who can provide constructive advice on improving your manuscript. In practice, this is likely to be your professor, who will provide comments and suggestions for improvement or may help you edit your manuscript. If you are lucky, you may have someone in your laboratory who can read your manuscript and provide feedback, such as a postdoc or senior graduate student. Indeed, it may be a good idea for graduate students and postdocs in a laboratory to agree to read each other's manuscripts, as this could help to improve the overall productivity of the group.

## Plagiarism

Plagiarism is a common problem in scientific research and should always be avoided. Plagiarism can be defined as a fraudulent representation of somebody else's work as one's own. For example, an author may copy large segments of another person's scientific manuscript and put it into their own manuscript. Many journals now use dedicated software to detect and quantify the amount of plagiarism in scientific manuscripts (*e.g.,* TURNITIN or IAUTHENTICATE). If the level of plagiarism is above some critical threshold (usually approximately 20–30%), then the article will be rejected. If the article has already been published, and then it is found to have plagiarized somebody else's work, it may be retracted by the publishers (officially removed from the journal). Having a retracted article gives an author a bad reputation and may make it more difficult for them to be published in the future. Consequently, it is always important to avoid plagiarizing other people's work.

Another issue is self-plagiarism. This is where an author repeats large sections of a previously published manuscript in their current manuscript. This is particularly challenging in the *Materials and Methods* section where an author may be using the same experimental protocols from a previous study, but they are now looking at new test samples. There are several ways to avoid self-plagiarism. First, you can rewrite the sections so there is less overlap between the original and new manuscripts. However, this is often unsatisfying because it can mean that your writing is less clear. Second, you can give a citation to the previous manuscript and then only include a short general statement of the methods used. For instance, you may say "The *in vitro* digestion of the samples was analyzed using a simulated gastrointestinal model that consists of mouth, stomach, and small intestine phases, as described in detail previously (McClements et al., 2023)." One of the drawbacks of this approach is that the reader must then find the cited manuscript, which takes time and energy.

The recent advent of artificial intelligence programs, such as ChatGPT, is creating new issues around plagiarism. How much of the writing of a scientific manuscript can be done by an AI program before it is considered to be plagiarism? Many academic publishers are developing guidelines and policies around this issue, and therefore it is advisable to check with the Author Guidelines of a journal if you are using an AI program to help with your writing. Typically, it is acceptable to use these programs to help write and check your work, but you are ultimately responsible for ensuring the information is accurate and the writing is appropriate.

## Concluding Remarks

The writing of scientific manuscripts is one of the most important things you do in your academic career because it's the way your findings are communicated to others and how others evaluate your scientific merit. Typically, the more publications and citations you have, the stronger your reputation in your research field will be, especially if these manuscripts are published in high-quality journals. In the case of graduate students, writing scientific manuscripts increases your chances of being awarded a PhD, reduces the time needed to complete your studies, and raises the possibility of you finding your next job. In this chapter, we have provided numerous tips on how to structure scientific manuscripts and how to write each section in a clear, concise, and engaging way. At the beginning of your career, writing a scientific manuscript can seem like a daunting prospect. However, as you do it more often, it becomes easier and can be one of the most rewarding parts of your scientific career. A well-written scientific manuscript is usually the culmination of many months or years of research and should therefore be written in a manner that will be most impactful. Consequently, it is critical to develop good scientific writing skills. These skills can be cultivated over time through practice, seeking feedback, and continuously striving to improve one's writing abilities.

**Key Points**
- Writing and publishing manuscripts is one of the most important aspects of being a successful academic.
- Initially, you should compile your data in a spreadsheet file, and then draw the various figures and tables that may appear in your manuscript.
- It is often useful to write the main sections of a manuscript in the following order: Materials and methods, Results and discussion, Introduction, Conclusion, Abstract, Title, Keywords.
- Authorship should only include those who have made a significant contribution to the manuscript, with the order of the authors depending on their relative contributions.
- Your manuscript should have a strong title that concisely describes the study, while avoiding hyperbole.

(continued)

(continued)

- Your *Abstract* should include concise information about the context, objectives, materials, methods, findings, importance, and implications of your research.
- Writing a good abstract and keywords helps other scientists find your work.
- Your *Introduction* section should include information about the context, background, objective, hypothesis, approach, novelty, and implications of your work. Ensure your citations of previous research are balanced, current, and relevant.
- Your *Materials and Methods* section should provide readers with a detailed description of how your study was carried out, so others can repeat it if desired.
- Your *Results and Discussion* section should clearly present, analyze, and interpret your data, and show how your findings relate to previous knowledge and advance the field.
- Your *Conclusions section* should discuss major findings, limitations, further research needs, context, and relevance.
- Reference managers are powerful tools for storing, citing, and formatting references.
- Review articles provide an overview of existing research on a specific topic. They are a good way of increasing your knowledge of the field, enhancing your reputation, and can be used as a literature review chapter in your thesis.
- When writing a manuscript, it is important to understand your audience, develop a strong narrative, be clear and precise, use correct grammar and spelling, be consistent in your writing style, and avoid plagiarism.

# Resources

Belcher WL (2019) Writing your journal article in twelve weeks, Second edition: a guide to academic publishing success. The University of Chicago Press, Chicago

Douglas MR, Bahr K, Olson R (2022) The narrative gym: for science graduate students and postdocs. Prairie Starfish Books, North Haven

Gastel B, Day RA (2022) How to write and publish a scientific paper, 9th edn. Greenwood, Santa Barbara

Heard SB (2022) The scientist's guide to writing: how to write more easily and effectively throughout your scientific career. Princeton University Press, Princeton

Olson R (2015) Houston, we have a narrative. Chicago Press, Chicago

Schimel J (2012) Writing science: how to write papers that get cited and proposals that get funded. Oxford University Press, Oxford

Springer (2023) Writing a journal manuscript. Springer, New York

Turabian KL (2018) Manual for writers of research papers, theses, and dissertations, 9th edn. The University of Chicago Press, Chicago, IL

# Chapter 7
# Submission and Revision: Getting Published

## Where to Publish Your Manuscript

Once you have written your scientific manuscript, you need to submit it to an appropriate journal. This can be a daunting task, as there are thousands of academic journals to choose from, each with different scopes, reputations, and formats. Therefore, you should decide the one that is most appropriate for your manuscript. Some of the most important factors to consider are highlighted in this section.

© The Author(s), under exclusive license to Springer Nature Switzerland AG 2024     153
D. J. McClements et al., *How to be a Successful Scientist*,
https://doi.org/10.1007/978-3-031-51402-9_7

## Subject Matter

Each journal only publishes articles that fall within its scope, which may range from very general to very specific. For instance, *Science* may publish articles from any area of science, provided they make important advances and appear impactful. The *Journal of the American Chemical Society* may publish articles from any area of chemistry, including physical, organic, inorganic, or analytical chemistry. The *Journal of Physical Chemistry* only publishes articles within physical chemistry. The journal *Langmuir* only publishes physical chemical research that focuses on colloidal particles and interfaces. Consequently, you can narrow down the potential journals to submit your manuscript by considering which subject area it best fits. Having said this, it is less important to precisely identify the most appropriate journal for your article these days, since most reputable scientific journals are online and can easily be searched using database search engines. As a result, potential readers are likely to find your article anyway.

## Manuscript Type

The academic journal you select also depends on whether your manuscript is an original research article or a review article. Some journals only publish review articles, whereas others publish a mixture of review and original research articles. If you have written a review article, and you believe that it is strong, it may be better to publish it in a journal that only publishes review articles because they often have a higher impact factor (see next section). Some journals only accept review articles that have been invited by the editor or editorial board (such as the *Annual Reviews* series). Consequently, you should carefully check the author instructions before submitting your article.

## Journal Reputation

The quality of your manuscript is partially judged by the reputation of the journal you publish it in, which is usually based on its *impact factor (IF)*. This quality index provides a measure of the average number of citations the articles published in a journal receive. It is calculated by dividing the total number of citations all the articles in a journal get in a specific year by the total number of articles published in the previous 2 years. The greater the number of citations of the manuscripts published in the journal, the higher its impact factor. The impact factor does not directly provide information about the quality of the editor, reviewers, or content of a journal. However, a journal with a high impact factor often attracts editors and reviewers with higher scientific reputations. The impact factor of journals varies greatly between disciplines, so a high value in one field may not be a high value in another field (Table 7.1). Articles in scientific disciplines where many people work, such as physics, chemistry, or biology, often receive more citations, which leads to higher journal impact factors. In contrast, articles in more specialized disciplines with fewer researchers, such as food or forestry science, typically receive far fewer citations, which leads to lower journal impact factors. Therefore, it is important not to compare the metrics of your research outputs with your peers/friends who might work in different areas. Ideally, you should aim to publish your research in the journal with the highest impact factor, which could be inside or outside your field, as this will increase the perceived quality of your article.

The reputation of scientific journals in a particular scientific field is also represented by placing them into four quartiles based on their impact factor: Q1 (top 25%), Q2 (25–50%), Q3 (50–75%), and Q4 (75–100%). It is advisable to publish in a Q1 journal whenever possible, as these have the strongest reputation. The quartile ranking of different scientific journals can be found online (www.scimagojr.com). Examples of quality indicators for selected academic journals in different scientific disciplines are shown in Table 7.1. As mentioned earlier, there can be large differences between different disciplines depending on the number of people working in the field, as well as the ease of publishing research in that field. Sometimes, it may be advisable to publish your work in a new journal that may not yet have an impact factor or has a low impact factor, provided that it will likely increase in the future. A publisher may introduce a new journal on a hot topic area. However, it typically takes at least 2 years for a journal to receive an impact factor, and in the first few years after that it may have a relatively low impact factor because people are hesitant to submit their manuscripts there due to the uncertainty of the journal's success. However, if you feel that the journal may have a high impact factor in the future, then it may be advantageous to submit a manuscript there. Moreover, the journals often give incentives to publish in them during the first 2 years, such as waiving publication fees.

## *Open Access*

Typically, academic journals work on either a subscription or open access model, which is based on the main way that the publisher makes money. Many journals use a subscription model, which means that libraries, companies, or individuals pay the publisher to subscribe to the journal and obtain access to the articles within it. Each journal typically costs a library or company a few thousand dollars for an annual subscription. The people working for the academic institution or company can then obtain access to the articles. In some cases, individuals may obtain access to a journal as part of their annual subscription fees to a scientific organization. For instance, members of the *American Society for the Advancement of Science* receive a copy of the journal *Science* every week as part of their subscription fees. Nevertheless, some journals that operate primarily on the subscription model still charge authors a publishing fee.

**Table 7.1** Some examples of the quartile ranking and impact factor of selected journals in different scientific disciplines

| Quartile | Biomedical Engineering | | Chemistry | | Physics and Astronomy | |
|---|---|---|---|---|---|---|
| | Journal | Impact Factor | Journal | Impact Factor | Journal | Impact Factor |
| Q1 | Nature Biotechnology | 68.2 | Chemical Reviews | 72.1 | Reviews of Modern Physics | 54.5 |
| Q1 | Nature Nanotechnology | 40.5 | Chemical Society Reviews | 60.1 | Living Reviews in Relativity | 40.4 |
| Q1 | Nature Biomedical Engineering | 29.2 | Nature Materials | 47.7 | Nature Reviews Physics | 36.3 |
| Q2 | Biosensors and Bioelectronics: X | 10.6 | Reactive & Functional Monomers | 5.0 | Results in Physics | 4.6 |
| Q2 | Journal of Biological Engineering | 6.2 | Journal of Photochemistry & Photobiology A: Chemistry | 5.1 | Journal of Non-Equilibrium Thermodynamics | 4.3 |
| Q2 | Nanotoxicology | 5.9 | Molecules | 4.9 | Chinese Journal of Physics | 4.0 |
| Q3 | Biomedical Microdevices | 3.8 | *Comptes Rendus Chimei* | 2.6 | International Journal of Geometric Methods in Modern Physics | 1.9 |
| Q3 | Biotechnology and Bioprocess Engineering | 3.4 | Journal of the Iranian Chemical Society | 2.3 | Indian Journal of Physics | 1.8 |
| Q3 | Biotechnology and Bioprocess Engineering | 3.4 | Supramolecular Chemistry | 2.2 | Chinese Physics B | 1.7 |
| Q4 | Technology and Healthcare | 1.2 | Journal of the Turkish Chemical Society, Section A: Chemistry | 0.8 | Brazilian Journal of Physics | 1.4 |
| Q4 | Biomedical Materials | 1.2 | Molecular Crystals and Liquid Crystals | 0.7 | International Journal of Fluid Power | 1.2 |
| Q4 | Micro and Nano Letters | 1.0 | Doklady Chemistry | 0.6 | Chalcogenide Letters | 0.9 |

Data accessed in mid-2023 using Scimago Journal and Country Rank (www.scimagojr.com). *Note*: There are large differences in the impact factors of journals in different fields

An increasing number of journals are operating on an open access model, which means that publishers obtain most of their revenue from authors who pay a fee to publish their manuscripts in the journal. In this case, the journal articles are made freely available to the public, which means that anyone can access them. Thus, the main advantage of publishing in an open access journal is that your work is accessible to a broader audience, which may increase its impact and citations. However, the main disadvantage is that you need to cover the fees associated with publishing your work. These fees can sometimes be included in government grants that fund scientific research projects. Therefore, if you or your professor has one of these grants, you may have sufficient funds to cover the publication costs. The fees to publish in subscription or open access journals can vary from a few hundred to a few thousand dollars, depending on the journal. Thus, if you work in a laboratory that is not supported by government funding or that publishes many manuscripts every year, you may not have enough funds to publish in journals that require a publishing fee. However, if you do not have sufficient funding, sometimes you can contact the publisher to get a reduction in fees. Moreover, there are still numerous reputable academic journals that work on a subscription model and do not charge a publishing fee. In addition, some open access journals occasionally invite scientists with a strong research reputation to publish in them for free because this increases the number of citations of the articles in the journal, thereby increasing the journal's impact factor and reputation. If you or your professor are one of these scientists and has an invitation, then you may be able to publish your manuscript in an open access journal for free, which can increase its impact.

## Predatory Publishers

When selecting an appropriate journal for your manuscript, you should beware of predatory publishers. These are companies that have mainly been established to make money from researchers by publishing their work for a fee. Of course, this is also true of many well-established publishers that have a good reputation. The main difference is that the editorial and reviewing processes are relatively weak for predatory publishers. The business model is to publish as many papers as possible to make a large profit from publication fees rather than to publish high-quality papers. As a result, many low-quality papers are accepted with little review. You should therefore avoid publishing in these journals. Many predatory journals bombard scientists with spam emails inviting them to submit their work for a discounted fee to make them sound more appealing. It is often advisable to set a filter on your email service to send these emails directly to your spam folder, so you don't have to deal with them. It should be noted that some publishers that were considered predatory in the past are trying to clean up their act as they become more established and put in place more rigorous editing and reviewing procedures. A useful list of predatory publishers and their journals was established by the librarian Jeffrey Beall, which is routinely maintained and updated (https://beallslist.net). If you are not sure about a particular publisher or journal, it is a good idea to refer to this list before submitting your manuscript.

## Manuscript Submission

Once you have identified a suitable journal and read the author instructions, then you are ready to submit your manuscript. You should go to the journal's website and click the appropriate "submit your article" link. The journal may then require you to log in to your online account with the publisher (if you don't have one, you will have to make one first). After you have logged in, there should be a link that allows you to submit your manuscript. Typically, you must select the type of manuscript it is, such as a research paper, review article, or short communication. You may also have to select whether the article is part of a special issue or not (usually, it is not), but sometimes you or your professor may get invited to submit an article to a special issue on a particular topic. You may then have to upload the different files associated with the submission. For many journals, you need to submit your manuscript, as well as several other documents, including a cover letter, graphical abstract, highlights, manuscript, figures, tables, author contribution statement, conflict of interest statement, and (sometimes) supplementary information. You will also need to input author details and suggest potential reviewers. The documents and information you need to provide depends somewhat on the journal, and can usually be found in the author instructions on the publisher's website. It is typically a good idea to prepare these documents before starting the submission process, but you can also prepare them as you go.

### *Cover Letter*

This letter should be addressed to the editor and should provide information about the title and authors of the manuscript. It may also include brief statements about the subject matter of the research, what was found, why it is important, why it is novel, and why it is suitable for publication in that particular journal (Fig. 7.1).

### *Graphical Abstract*

The graphical abstract, which is sometimes referred to as the table of contents image, provides a visual representation of your manuscript's main findings. It is designed to capture the reader's attention and provide a summary of the main contents of the article. It is typically displayed on the journal's website as part of the table of contents of the issue your article appears in. A visually attractive and informative graphical abstract can increase the readership and impact of your research. The journal often gives specific instructions on the format of a graphical abstract that should be followed, such as its dimensions and resolution. The graphical abstract should be original, created by the authors, and not published elsewhere. It

David Julian McClements
Department of Food Science
University of Massachusetts
Amherst, MA 01003

June 1, 2023

Dear Editor

We would like to submit a manuscript entitled "*Title of article*" by McClements and co-workers to *Journal Name* as an original research article. The objective of this research was to…. We found that …. We believe that this research is original because….

We look forward to hearing from you in due course.

Yours Sincerely,

David Julian McClements
Distinguished Professor
413 545 2275; mcclements@foodsci.umass.edu

**Fig. 7.1** A template for a cover letter that should accompany a scientific manuscript when it is submitted to a journal

should contain text and diagrams that are easy to read (not too small). A graphical abstract should not include an image of any person. Ideally, the graphical abstract should be different from the figures included in the manuscript. Graphical abstracts are usually prepared using specialized presentation or drawing software programs such as PowerPoint, Blender, or Photoshop. An example of a graphical abstract is shown in Fig. 7.2.

## *Manuscript*

The document containing your manuscript is the most important part of your submission. It typically contains all the text associated with your manuscript and is prepared using a word processing program (like Microsoft Word). This document must be formatted according to the requirements of the journal, including the abstract, sections, headings, and reference styles. Some journals require you to convert your manuscript file to PDF format before submission.

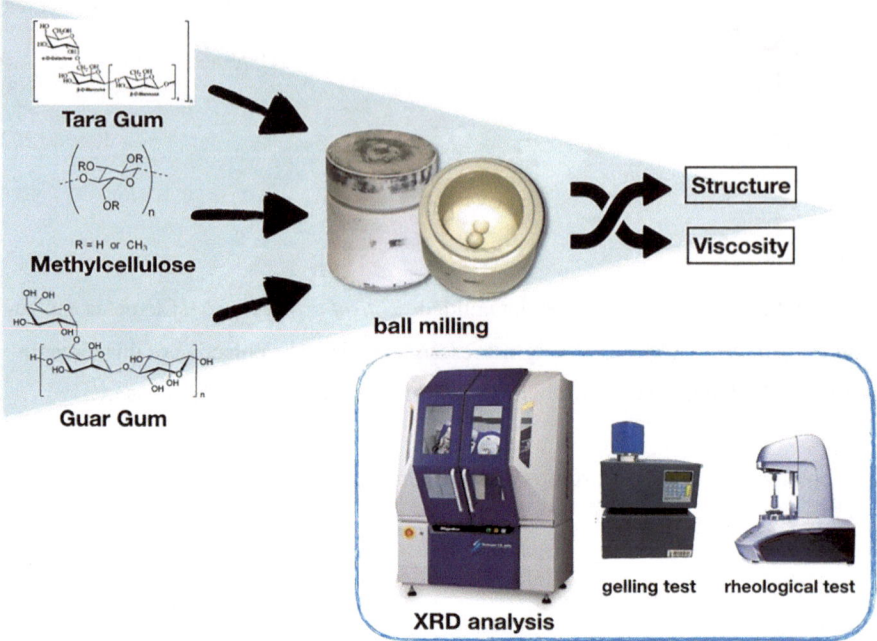

**Fig. 7.2** Example of a graphical abstract that is uploaded with a manuscript during the submission process. This example is from a research article on the effects of ball milling on the functional properties of hydrocolloids. (Nuvoli et al. 2020) (Creative Commons CC BY 4.0 license)

## *Figures and Tables*

Some journals allow the figures and tables to be embedded within the manuscript file where they would appear in the published article. This makes it easier for the reviewers to read and review your manuscript since the figures and tables are near the place where they are discussed in the text. However, some journals require figures and tables be included at the end of a manuscript or be uploaded as separate files. There may be formatting requirements for the figures and tables, such as the dimensions and digital resolution of any images, which can be found in the author instructions for the journal.

## *Supplementary Information*

If a journal allows you to include supplementary information to the main article, such as additional text, figures, tables, equations, or code (see Chap. 6), then you could upload this information as a single file during the submission process. Readers of your manuscript who are interested in this material can usually access it online from the publisher's website.

## *Author Contribution Statement*

Many journals now require authors to provide a CRediT (contributor roles taxonomy) statement when they submit their manuscripts. This statement allows the authors to provide information about the various contributions that each person made to the study. Typically, the corresponding author is responsible for ensuring that these descriptions are accurate and have been agreed upon by all the other authors. Each author may have more than one CRediT term associated with their name, but it is not necessary to use all the available terms. Examples of the main terms used to describe author contributions to a manuscript are listed in Table 7.2, whereas an example of a credit statement is shown in Fig. 7.3.

**Table 7.2** Terms used to describe different author contributions to a manuscript in the CRediT statement needed by some journals

| Term | Definition |
|---|---|
| Conceptualization | The author contributed to the initiation and/or development of the idea that the research is based on |
| Data curation | The author contributed to the design, establishment, maintenance, and accessibility of any large databases produced in the study |
| Formal analysis | The author contributed to the analysis and synthesis of the data using statistical, mathematical, or computational methods |
| Funding acquisition | The author helped to secure the funding that supported the research |
| Investigation | The author was involved in performing the research, such as carrying out experiments or collecting data from observations or surveys |
| Methodology | The author was involved in developing the experimental methods or protocols used in the study |
| Project administration | The author was responsible for management and coordination of the research project |
| Resources | The author provided key resources for the study, such as materials, reagents, samples, test animals, computer models, *etc.* |
| Software | The author helped to develop or employ computer software that was key to the success of the project |
| Supervision | The author provided oversight and leadership to the team who carried out the study |
| Validation | The author was responsible for validation of the materials or methods used in the study |
| Visualization | The author was responsible for preparing figures or graphs that appear in the scientific manuscript |
| Writing-original draft | The author contributed to the writing of the original draft of the scientific manuscript |
| Writing-review and editing | The author contributed to reviewing and editing the scientific manuscript |

Based on the terms developed by Brand et al. (2015)

---

**Credit Author Statement**

*Isobelle Farrell McClements*: Conceptualization; Methodology; Investigation; Formal analysis; Visualization; Writing - original draft. *Jake   McClements*: Conceptualization; Formal analysis; Writing - original draft. *David Julian McClements*: Conceptualization; Supervision; Writing - review & editing.

---

**Fig. 7.3** Example of a CRediT author statement that some academic journals require when you are submitting your manuscript

## Selection of Reviewers

Many journals require you to provide the names and contact information of several people (typically 2–5) with the appropriate expertise to review your manuscript, which the journal editor may or may not contact. You may therefore have to do some research to identify potential reviewers. These should not be people who you have a close connection with, or that you have published articles with recently. They should be people who work in a scientific area that is related to the topic of your manuscript. You or your professor may know other research groups that work in a similar area to you, so you could select a person from one of these groups. Alternatively, you can perform a literature search using appropriate keywords to find other people who have recently published in the same area. Typically, it is advisable to select reviewers whose work you have cited in your manuscript.

## Funding Information

Most journals require you to provide information about the organization that funded your research, such as a government program, industrial sponsor, or nonprofit organization. This may be included as a separate section at the end of your manuscript. However, you may also have to enter it as part of a drop-down menu on the publisher's website during the submission process.

## Conflict of Interest

Many journals require you to submit a "Conflict of Interest" statement with your manuscript. Sometimes this can be included as part of the main article but in other cases, it may have to be uploaded as a separate file. An example of a "Conflict of Interest" statement was provided in Chap. 6.

## *Final Submission*

Once you have uploaded all the required information on the publisher's website, you are ready to submit your manuscript. Typically, the website converts all the files you have uploaded into a single PDF file. You should then carefully review this file to ensure all the information is correct before submitting the article. Once you click the final submission link, you will get a confirmation that your article has been successfully submitted. Before logging off from the journal website, you should save a copy of your manuscript PDF file. This file can be useful when you receive feedback on your manuscript from the reviewers, *e.g.,* to locate the line numbers in the manuscript where any changes are needed.

Once you have submitted your manuscript, you have to wait to see whether your article will be accepted or rejected by the journal. There are several possibilities at this stage:

- *Sent back to the author*: The journal identifies some mistake in your manuscript, such as incorrect formatting or a missing file, and will send it back to you to correct. You will then have to resubmit the manuscript. This will delay the publication of your research, so it is important to carefully upload all the required files in the correct format during the initial submission process.
- *Rejected by the editor*: The editor may feel that the manuscript does not fit the scope of the journal, or they may think the quality of the manuscript does not meet the standards of the journal. They may then reject the manuscript without sending it out for review. Sometimes, they may suggest other journals that are more suitable for your manuscript.
- *Sent out to review*: If the editor thinks your manuscript is on a suitable topic and of suitable quality, they will send it out for review. Typically, they send it to two or more reviewers who are expected to provide critical feedback on your manuscript. The reviewers are asked to judge the scientific merit of the manuscript based on several criteria, including the appropriateness of the subject matter to the journal, the originality and significance of the findings, the appropriateness of the methodology, the strength of the results and conclusions, and the quality of the writing, figures, tables, and diagrams (Fig. 7.4). It can take anything from a few weeks to a few months for your manuscript to be reviewed. The review process can sometimes be quite long because it is increasingly difficult for editors to find individuals who have the time and willingness to review articles.

## Manuscript Revision

After your manuscript has been reviewed, the journal editor will provide a recommendation on its suitability for publication based on the reviewers' comments and suggestions. There are several possible outcomes of the editor's decision:

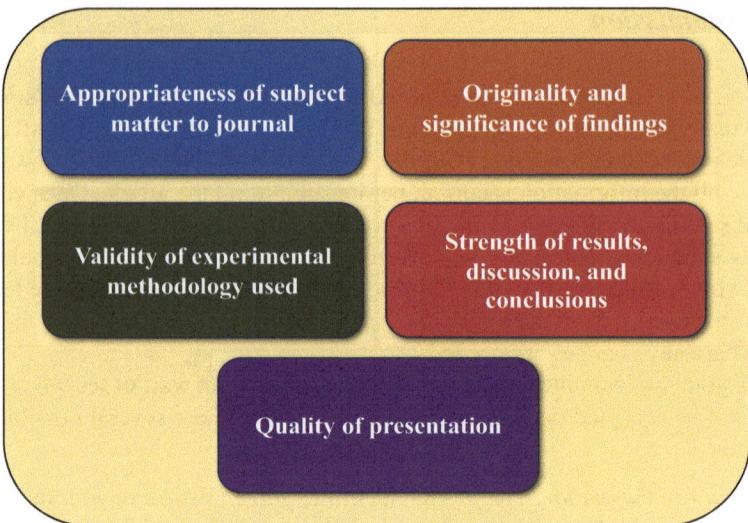

**Fig. 7.4** Reviewers are asked to judge the merit of scientific manuscripts submitted to academic journals based on several criteria

- *Publish as is:* In some very rare cases, an editor may recommend that your article be published in its original format, with no revisions required.
- *Publish with minor revisions*: An editor may state that your article is suitable for publication but requires some minor revisions based on the reviewers' comments before it can be accepted. In this case, you must make the required revisions and then resubmit the manuscript. The editor may then check to see whether you have made all the required changes, and if you have, accept the manuscript. If you have not made the changes to the editor's satisfaction, they may send the manuscript back to you for further revision.
- *Publish with major revisions*: An editor may state that your article might be suitable for publication, but it requires extensive revisions based on the reviewers' comments before it can be considered again. Therefore, you must make the required revisions and then resubmit the manuscript. In this case, the editor may send your article back to the reviewers so they can determine whether you have made sufficient changes to make it suitable for publication. This process may be repeated several times if you do not fully address the reviewers' concerns in the first or second revision.
- *Reject*: Based on the reviewers' rankings and comments, an editor may simply reject your article. This may occur because the subject matter is deemed to be inappropriate for the journal, the study is not novel, the research findings are trivial, the manuscript is poorly written, or the results are not interpreted appropriately. If you feel that the reviewers did not provide a reasonable review of your manuscript (which sometimes happens), you may be able to appeal to the editor. You would need to give good reasons why you think the review is unfair. The

editor may then ask one or more additional reviewers to look at it. Alternatively, you could revise the manuscript based on the editor's and reviewers' comments and submit it to another journal.

In the following sections, we discuss some of the nuts and bolts of effectively responding to reviewers' comments and suggestions, thereby increasing the chances of your manuscript being accepted.

## The "Responses to Reviewers' Comments" Letter

Typically, you need to provide a "Responses to Reviewers' Comments" letter with your revised manuscript. This letter contains your detailed responses to each of the comments and suggestions made by the reviewers. This process can be simplified by using a standardized format to prepare the response letter. You want to make this process simple for the editors and reviewers (they are typically very busy people) and to recognize the time and effort they have put into handling and reviewing your manuscript. Reviewing scientific manuscripts can be a time-consuming and laborious process that does not lead to any financial reward or official recognition but is critical to the advancement of science. It helps to weed out poor work and to improve the quality and impact of published research.

When writing your response letter, you should start by providing information about the manuscript so that the editor can easily identify which article they are considering. For example:

*Manuscript Number*: FOODHYD-D-22-03752
*Title*: Modification of textural attributes of potato protein gels using salts, polysaccharides, and transglutaminase: Development of plant-based foods
*Corresponding Author*: David Julian McClements
*Journal*: Food Hydrocolloids

You should then include a statement that acknowledges the efforts of the reviewers in providing constructive feedback about your manuscript.

**Responses to Reviewer Comments**
– We thank the reviewers for their constructive comments and suggestions on our manuscript. We have revised the manuscript accordingly and included a detailed list of responses below. After these improvements, we hope our manuscript will now be suitable for publication.

It is then important to address all the comments and suggestions made by the reviewers in a respectful and comprehensive fashion. Reviewers may make general comments about your manuscript, such as the importance of the subject matter, the novelty of the study, the quality of the research, and the overall quality of the writing. They may also provide more specific comments on particular parts of the manuscript where they believe information should be added, removed, or revised to make the manuscript better. Typically, the comments and suggestions made by reviewers require different actions:

- *Agree (no further experiments):* In some cases, you agree with a reviewer's comment, and it is relatively simple to make the requested change to your manuscript without having to perform additional experiments or observations. In these cases, you simply make the revisions suggested and describe what you have changed below the reviewer's comment.
- *Agree (further experiments)*: In some cases, you agree with the reviewer's comments but need to perform more experiments or observations to provide some additional information requested by the reviewer. In these cases, you should carry out the additional experiments or observations, revise the manuscript, and then describe what you did in the response to the reviewer's comment.
- *Disagree:* In some cases, you disagree with the reviewer's comment because you think it is inappropriate or that they missed something that was already included in the original manuscript. In these cases, you must provide a strong argument for not making the changes suggested by the reviewer.

Some examples of a reviewer's comments and suggestions, as well as potential responses to them, are given here as a guide. The reviewer's comments are given in plain text, whereas the responses are given in italics.

**Responses to Reviewer 1**
**General comments:**

This study was on the utilization of plant-based proteins and polysaccharides to create meat analogs. It is on an important subject and is generally well-written. The authors have provided new information on the physical chemistry of creating meat analogs that may be useful to food manuscripts. I believe it can be published with minor changes, as highlighted below. However, the authors should check the grammar throughout the manuscript, as there are a number of typographical errors.

– *We thank the reviewer for their positive comments on our manuscript. As requested, we have revised the manuscript throughout to correct any grammatical errors.*

(continued)

(continued)

**Specific comments:**

Line 89: The word "temporory" is spelled incorrectly

– *As requested, we have changed "temporory" to "temporary" in the revised manuscript.*

Line 123: The authors need to provide more information about how the protein solutions were prepared.

– *As requested, we have revised the materials and methods section to provide more detailed information about how the protein solutions were prepared.*

Line 456: The authors did not measure the electrical properties of the proteins and polysaccharides used in the study. This information is critical for understanding how they interact with each other due to electrostatic interactions.

– *We thank the reviewer for this suggestion. In response, we have carried out additional experiments to measure the electrical properties of the proteins and polysaccharides using particle electrophoresis. These are included in Figure 2 and discussed in Section 3.3 of the revised manuscript.*

Line 473: The authors did not consider the importance of hydrophobic interactions in their analysis.

– *We agree with the reviewer that hydrophobic interactions are important. Actually, we discussed the importance of these interactions in Section 3.4 of the original manuscript. In response to the reviewer's comment, we have expanded this discussion.*

Once you have responded to all the reviewers' comments and suggestions and revised your manuscript accordingly, you are ready to submit the revised version to the journal. Some journals require that you submit two revised versions of the manuscript: (i) a *marked version* highlighting all the changes you made, which can be done by highlighting the changes in red or using the "Track Changes" function of a word processing program; (ii) an *unmarked version* containing all the changes. The marked version is useful for the reviewers because it allows them to quickly locate the parts of the manuscript where any changes were made. The unmarked version is useful for the publisher since it can simply be typeset once the manuscript is accepted. Once you have prepared these documents, you can then log on to the journal's website, locate your article, and start a revision. Typically, you need to submit revised versions of all the files you have modified, as well as a cover letter with the detailed responses to reviewers.

## *Comments on Reviewers*

As mentioned earlier, the people who review scientific manuscripts to ensure they are suitable for publication play a critical role in science. Reviewers have different personalities, temperaments, and expertise, which influences their review of your manuscript.

Some reviewers are polite, positive, and constructive, whereas others can be impolite, negative, and disparaging. It is important not to take any negative reviews of your work personally. Wherever possible, use the comments of the reviewers to improve the quality of your work, and provide responses that are courteous and address all the issues raised by the reviewers. The reviewers have spent considerable time reading your work, so you should take their comments seriously. If you do not agree with their comments, then you can politely explain why, giving convincing evidence and reasons to support your argument. If you think a reviewer has been unfair or overly critical of your work, you may point this out to the editor and provide strong reasons why you think the reviewer has not given your manuscript a fair review. If the editor agrees with you, they may take this into account when deciding whether to accept or reject your manuscript.

## Manuscript Proofreading

After your paper has been accepted, it will be sent to the publisher who will put it into the format required for final publication. The manuscript will then be sent back to you for one final check, which is known as proofreading. The publisher may have some queries on your manuscript that you need to respond to, such as checking the correct spelling and affiliations of the authors, checking the correct location of the figures and tables, and highlighting citations that may be missing, incorrect, or duplicated. You should also carefully read through the typeset version of your manuscript to check for any grammatical errors and to be sure that all the tables, figures, and headings are correct and numbered properly. Typically, you cannot make large-escale changes to your manuscript at this stage, such as including new paragraphs, sections, figures, or tables.

After you have carefully proofread your manuscript and answered all the queries, you can submit the final version. It may then take a few weeks or months for the official version of your manuscript to be published in an issue of a journal, which may be online or in press.

## Manuscript Promotion

Once your manuscript has been officially published, you may want to start promoting it. For example, you could create a post about your new article on social media outlets that people working in your field might read. Many journals now provide a link to recently published scientific manuscripts that allow people to access them for free. However, there may only be a limited number of times the article can be downloaded for free (often 50), or there may only be a limited timeframe after publication that the article can be downloaded for free (often 50 days). If the journal sends you an email with this kind of link to your manuscript, then you are highly advised to post it on an appropriate social media account where other scientists in your field may be able to access and download it (such as LinkedIn). This may increase the impact and citations of your research. If your manuscript represents an important scientific advancement or it may be interesting to the general public, then you can inform the Press or Media Office at your university. They may then create a feature based on your article that they circulate to media outlets, thereby potentially increasing the visibility of your work.

## Concluding Remarks

The reviewing and revision of scientific manuscripts are major factors contributing to the progress of science and technology. Reviewers help to identify errors in the design, performance, or interpretation of a scientific study that the original researchers were unaware of, as well as provide new ideas for additional experiments or interpretations that can help improve the researcher's work. The reviewers may also be able to provide suggestions on how the results can be presented in a more compelling and understandable fashion, which could increase the impact of your research. Reviewers are often the unsung heroes of science – receiving no financial rewards or broader recognition. The results of the submission and revision process can be highly rewarding (if your manuscript gets accepted) or highly disparaging (if your manuscript gets rejected). It is good advice to develop a respectful and constructive attitude when responding to reviewers' comments. You should think of the comments as free advice from experts on how to improve your manuscript, rather than as criticisms of you or your work. If you are a graduate student, the feedback provided by the reviewers will result in much stronger thesis chapters that are less likely to be adversely critiqued by your examiners. It would be difficult to get this level of expert proofreading and critique of your work, without submitting it to a journal. If your manuscript does get rejected by one journal, do not take it personally, this is a normal part of the scientific process. If you can, you should carry out any additional research needed, revise your manuscript, and then submit it to another journal. If your research has merit, then it should eventually get published, thereby advancing our understanding of the world.

**Key Points**
- When choosing a journal to publish in, you should consider the subject matter, manuscript type (original research or review), and journal reputation.
- Journal reputation is indicated by the *impact factor*, which is based on the average number of citations each article receives.
- Typically, journals work on a subscription and/or open access model. Many journals require authors to pay a publication fee. Open access journals are freely available to other scientists and therefore tend to be more highly cited.
- It is important to avoid predatory publishers who may publish low-quality papers to gain profit. There is an online database to check whether a publisher is considered predatory.
- When submitting a manuscript, you typically need to also include a cover letter, graphical abstract, highlights, figures, tables, author contribution statement, and possibly supplementary information. You also need to provide author details and potential reviewers.
- The cover letter should be addressed to the editor and provide concise information about the title, authors, and content of the manuscript, as well as its relevance to the journal.
- The graphical abstract provides a visual representation of the manuscript's main findings. It should capture the reader's attention while providing a summary of the main contents.
- The author contribution statement provides information about the different contributions made by each author to the final manuscript.
- After your manuscript is reviewed, there are several possible outcomes. The paper may be published as is, published with minor revisions, published with major revisions, or rejected.
- If a reviewer provides suggestions for revision to your manuscript, you should provide a "Responses to Reviewers' Comments" letter with your revised manuscript.
- You may agree with a reviewer's comment and make the requested change (which may involve further experimentation) or disagree and then politely explain why.
- After your paper has been accepted, it will be sent to the publisher who will put it into the correct format for publication. It will then be returned to you for a final proofreading.
- After publication, it is advisable to promote your manuscript on a social media platform such as LinkedIn where other scientists can view it, as this may increase its visibility and citations.

# Resources

Brand A, Allen L, Altman M, Hlava M, Scott J (2015) Beyond authorship: attribution, contribution, collaboration, and credit. Learn Publish 28(2):151–155

Gastel B, Day RA (2022) How to write and publish a scientific paper, 9th edn. Greenwood, Santa Barbara

Heard SB (2022) The scientist's guide to writing: how to write more easily and effectively throughout your scientific career. Princeton University Press, Princeton

Nuvoli L, Conte P, Garroni S, Farina V, Piga A, Fadda C (2020) Study of the effects induced by ball milling treatment on different types of hydrocolloids in a corn starch–rice flour system. Foods 9(4):517

# Chapter 8
# Conferences and Networking: Getting to Know the Science and the People Behind It

## What Is a Conference?

Conferences bring together individuals to share their research findings, network with peers, and stay up to date on advances within their respective fields. They are an integral part of academic life and are often held at different locations each year. Attendees typically include individuals from academia, ranging from graduate

© The Author(s), under exclusive license to Springer Nature Switzerland AG 2024          173
D. J. McClements et al., *How to be a Successful Scientist*,
https://doi.org/10.1007/978-3-031-51402-9_8

students to professors, as well as members of industry, government agencies, and clinical organizations. Conferences usually focus on specific research areas to ensure that the content is relevant to attendees. However, the scope of these areas can vary significantly, ranging from highly multidisciplinary meetings to those that focus on specialized topics. Similarly, the size of conferences can range from international events that attract tens of thousands of attendees to smaller gatherings with fewer than 100 delegates. In general, conferences are advertised well in advance, and delegates must register and pay a fee to attend. Delegates may apply to present their research findings as a poster or oral presentation, which helps to disseminate their research. This chapter offers a practical guide on all aspects of conferences and provides insights into how graduate students and other scientists can make the most of these important events.

## Why Are Conferences Important?

As you read this chapter, you may be thinking: "Are conferences really that important when I can stay up to date with the latest research and connect with peers online?" The answer to that question is a resounding "yes", as conferences are an important aspect of the scientific process for many reasons:

- *Attaining presentation experience:* Conferences provide graduate students and other scientists with valuable opportunities to showcase their research through poster or oral presentations. While it may seem daunting, presenting research at conferences will build confidence and improve communication skills, which will be valuable regardless of your future career path. Additionally, presenting at conferences can establish your reputation within the research field and enhance your CV (Chap. 11). Typically, graduate students are encouraged to aim for *at least* one poster and one oral presentation at conferences during their PhDs.
- *Receiving Feedback*: Another valuable aspect of presenting at conferences is the opportunity to receive feedback on your research from experts in the field. Although it may be nerve-wracking to answer questions in front of an audience, the feedback gathered can provide fresh perspectives and insights that may be pivotal in making breakthroughs in your research. Additionally, answering questions about your research is excellent practice for when you come to defend your final thesis in front of professors.
- *Generating Ideas:* Beyond receiving feedback on your own research, conferences offer the chance to stay informed on advances within your field. Research presented at conferences is often very recent and can be unpublished. Consequently, this allows attendees to remain up to date with cutting-edge research. Additionally, unlike reading papers, you can ask questions about the research in real-time to gain a deeper understanding of the work. This can provide valuable insights to guide your own research and generate new ideas for experiments.

- *Networking*: Networking is a particularly important aspect of attending conferences. It provides the opportunity to build relationships with a diverse range of individuals and exchange valuable ideas about your research. Additionally, networking can facilitate collaborations with individuals possessing unique skills, knowledge, and equipment not typically available to you. These collaborative relationships can result in exciting joint projects that can generate important results for publications, funding applications, or theses. Moreover, networking at conferences is an excellent way to discover new funding and employment opportunities, as well as to make a positive impression on influential individuals within your field.
- *Publishing:* Some scientific organizations publish conference proceedings, which include manuscripts based on oral or poster presentations at the conference. Papers published in these proceedings are typically shorter and undergo less rigorous review than those published in scientific journals, which can reduce their credibility. However, they still help to disseminate your research and build your publication record. Therefore, this may be a viable option if you are struggling to publish your work in a journal or do not have enough results for a full article. Additionally, some journals release special editions associated with conferences, to which delegates who presented their work are invited to submit manuscripts. However, these manuscripts must still meet the journal's standards and undergo a rigorous review process.
- *Life Experience:* Attending conferences is not only about work. They also provide fantastic opportunities to explore new places, meet diverse groups of people, and learn about different cultures. During my PhD (JM), I was fortunate enough to attend conferences in various locations across the UK, as well as in Madrid and Rome. While these experiences helped me academically, they were also fun and provided valuable life experiences.

## What Happens at a Conference?

So now you're sold on why conferences are important, interesting, and even fun, we can explore what typically happens at these events. Arriving at your first conference can be quite intimidating, as you may be unsure where to go, what to do, or who to talk to. However, they generally follow a similar format, so you will soon become adept at navigating them. The first step is registering your attendance and collecting your name tag and information pack. We suggest wearing your name tag throughout the conference so other attendees know who you are and where you work. Your information pack usually contains a program that serves as your guide for the conference. However, to make the most of your experience, it's best to read the program online beforehand (more on this later!). The registration desk is typically located at the conference entrance, making it easy to find and a good place to ask any practical questions.

**Table 8.1**  A typical schedule for the first morning of a larger conference

| 9:00 | Welcome | | | |
|------|---------|---|---|---|
| 9:30 | Plenary Talk | | | |
| 10:30 | Break | | | |
| | **Session 1** | **Session 2** | **Session 3** | **Session 4** |
| 11:00 | Keynote Talk | Keynote Talk | Keynote Talk | Keynote Talk |
| 11:30 | Oral Presentation | Oral Presentation | Oral Presentation | Oral Presentation |
| 11:50 | Oral Presentation | Oral Presentation | Oral Presentation | Oral Presentation |
| 12:10 | Oral Presentation | Oral Presentation | Oral Presentation | Oral Presentation |
| 12:30 | Lunch and poster session | | | |

Once you have registered, the conference will begin with a brief welcome presentation from the organizers. Your time will then typically be divided between attending presentations and networking. At smaller conferences, presentations are usually held in a single session intended for all attendees. However, larger conferences can have multiple parallel sessions focused on different research topics. This means you can only attend a portion of the total presentations, so it's important to choose carefully. A typical schedule for the first morning of a larger conference with multiple parallel sessions is presented in Table 8.1. Conference presentations generally consist of a talk accompanied by a slideshow that lasts about 75–90% of the allocated time slot. The remaining time is reserved for audience questions. A session chair is responsible for introducing the speakers, keeping the presentations on schedule, and facilitating audience questions. Organizing committees typically classify conference presentations into the following categories:

- *Plenary lectures*: These presentations are an important part of conferences as they are scheduled when no other activities are taking place to allow all delegates to attend. The plenary speakers are invited by the organizing committee and usually have an outstanding reputation within their field. Consequently, these lectures will be of interest to most attendees. Plenary lectures will usually be the longest conference presentations, lasting 45 minutes to an hour. Depending on the conference type, plenary sessions may also include a panel discussion or other types of presentations.
- *Keynote lectures:* These presentations are also given by distinguished speakers who the organizing committee invites due to their strong reputation. However, unlike plenary lectures, keynote presentations are focused on specific themes and can take place when other activities are scheduled, depending on the conference size. Typically, keynote lectures are given at the start of a session to help establish the session's theme, and they usually last about 30–45 minutes. They are of high interest to many attendees and can provide valuable insights into current research.
- *Oral presentations:* In a conference session, each keynote lecture is typically followed by several oral presentations that relate to the session's theme. These presentations are shorter, usually about 15–30 minutes, and the organizing committee may not invite speakers. Instead, attendees can apply to give an oral

presentation on a specific theme by submitting an abstract outlining their research. Oral presentations can be given by any attendee and offer an excellent opportunity for graduate students to gather valuable presentation experience and establish their reputation within the field.

- *Poster sessions:* Before attending a conference, delegates can prepare posters that provide an overview of their research. These posters are displayed on boards in a specific location for some or all of the conference. Furthermore, there are often dedicated poster sessions, typically accompanied by refreshments, where attendees are invited to peruse the posters for a couple of hours while the presenter is there. The presenters generally stand by their poster to answer questions or exchange ideas with interested delegates. Posters are an excellent way to present your work to a large audience and have in-depth discussions with peers. Although any attendee can present them, they are mostly produced by graduate students and early career researchers. Similar to oral presentations, delegates must submit an abstract to present a poster. At the end of a conference, prizes are often awarded for the best posters, which can be significant and make an excellent addition to your CV!
- *Panel discussions:* These sessions typically involve a panel of experts discussing a topic relevant to the conference. They aim to encourage thought-provoking conversations and provide the audience with new perspectives on various topics. Different formats can be used for these sessions, including discussions between the panelists, short speeches, or fielding questions from the audience. While less common than conventional presentations, panel discussions can be highly informative and engaging for attendees.
- *Tutorials/workshops:* These are distinct from typical conference presentations, as they offer training or guidance to attendees on specific topics rather than presenting research findings. For instance, there may be sessions on understanding experimental techniques or preparing manuscripts for top journals. These presentations are mostly attended by graduate students or early career researchers and provide a great opportunity to broaden your knowledge base. Some conferences also offer dedicated summer schools or training courses on specific topics that occur before the main conference begins.

## Choosing a Conference

A large number of conferences take place every year, making it difficult to know which one is the best to attend. An excellent place to start is usually speaking to your lab mates, as the more senior ones will often have attended conferences that may be relevant to you. Therefore, they can provide some useful recommendations. Additionally, if you and your lab mates work in a similar research area, it can be nice to attend conferences together. You should also look online to find conferences within your research field. Although the full program is unlikely to be published far in advance of the conference date, the session topics and plenary speakers should be listed, so you can see if the topic is relevant to your work. Ideally, plenary speakers should be prominent contributors to your research field. You can find conferences online using a search engine or set up alerts for conferences within your field using a database website such as *Conference Alert* or *All Conference Alert*. Social media is another great place to find relevant conferences, particularly LinkedIn, ResearchGate, or X. Posts are often used to advertise the conference by the organizers or from members of your network that will be presenting there. Large publishers or scientific organizations, such as the *American Chemical Society*, often hold annual conferences, so you can check the websites of relevant organizations and publishers within your field to see if they do this. If you are a graduate student or postdoc, once you have found a conference relevant to your field, you can discuss the potential of attending it with your supervisor.

It is important to avoid predatory conferences. These are unethical conferences that are organized with the sole aim of making money from registration fees rather than advancing scientific knowledge. Therefore, they often lack rigorous peer review, genuine scholarly intent, and proper organization. You may frequently receive invitations to present at predatory conferences, which might seem out of the blue and have little relevance to your field. Furthermore, the emails are often poorly written, and the corresponding websites are low quality and unprofessional. Legitime conferences will generally not spam researchers with requests to attend/present, and their websites will often be of high quality. If you are unsure about a potentially predatory conference, investigate online and speak to your lab mates and supervisor before submitting an abstract or paying a registration fee.

## Practical Considerations

In an ideal world, every PhD student could attend the most prestigious and relevant conferences within their research field (Fig. 8.1). However, this is typically not the case, as many practical considerations can dictate whether you can attend a conference. Primarily, your supervisor must have the funding available to cover your registration fees, travel, accommodation, and subsistence. For large international conferences, the total cost may be quite substantial; therefore, it may not be possible for your supervisor to fund you. However, some departments and universities do offer funding to fully or partially cover the costs of PhD students to attend conferences. Typically, you may have to apply for a small travel grant to receive this kind of funding, which often have favorable success rates. Furthermore, you might consider volunteering at a conference, which might then waive your registration fees and might even provide you with some funds to cover travel and hotel costs. For conferences held by scientific organizations, registration fees are often significantly lower for members, so it may be worth joining the organization before registering.

Given the cost of conferences, it is important to be strategic when choosing which ones to attend. For example, you may only get the opportunity to go to one major international conference during your PhD, so it's generally better to attend

**Fig. 8.1** Presenting your work at scientific conferences is important for increasing your visibility and the impact of your research

this toward the end of your studies. This is because you will have more significant results and thus are more likely to get an accepted oral presentation, which will look more favorable on your CV and boost your scientific reputation (Chap. 12). Moreover, through networking, you may be able to find opportunities for your next position after completion of your PhD. Identifying multiple conferences with different costs and exploring these options with your supervisor is a good idea. For example, some conferences are held locally or aimed at early career researchers, which significantly reduces their costs. Furthermore, online conferences can be an excellent way to present your research to a large audience with no associated costs. Although these events may not be as prestigious as major conferences, they still offer an excellent opportunity to present your research, grow your reputation, build your network, and hone your presentation skills.

Another practical consideration is whether attending the conference is worth the time away from your graduate studies. While conferences are important for PhD students, and we strongly encourage you to attend them, sometimes they result in significant time away during a vital part of your graduate studies. For example, if you have a tight deadline to finish your experimental work or submit your thesis, then it is probably not a good time to attend a conference. This can be hard to predict as you will register in advance, but it's worth considering. You may also want to think about what else the conference can offer you. For instance, some conferences hold workshops, summer schools, and training courses, which may be very useful and are sometimes included in the registration fees. Moreover, you might be able to publish your work in a special edition of a journal associated with the conference. Some conferences also have career fairs that can help you identify a future employer. Therefore, understanding what the conference can offer you is important to get the most out of it. Although you shouldn't decide whether you want to attend a conference based on its location, it's often a bonus if it is held somewhere you would like to visit!.

## Before the Conference

Once you have decided which conference you want to attend and confirmed the costs with your supervisor, there are several tasks you need to complete. First, you must register your attendance on the conference website. This typically involves filling in your personal details, paying the registration fee, including any add-ons such as conference dinners or training courses, and submitting an abstract for a poster or oral presentation. Registering as early as possible is recommended, as the fee often increases closer to the conference date. Likewise, booking accommodation and travel is recommended soon after registration to enable you to get the best possible deal. These bookings may be made through your university or a university-associated travel agency. However, sometimes you might need to book these directly and then claim the money back. This can be financially difficult for PhD students, so if it's not possible for you, then discuss it with your supervisor and a solution should be possible. Your university may have accommodation specifications linked to their insurance policies, such as hotels rather than private residences (such as Airbnb). It is therefore essential to confirm this before booking anything. After booking is complete, you will likely need to submit a travel authorization to your university, which will include your travel details and the funding source used to cover the costs.

## *Submitting an Abstract: Oral and Poster Presentations*

For every conference you attend, we recommend submitting a poster or oral presentation abstract, as these make great additions to your CV, help build your skills and confidence, and grow your reputation. Deciding which presentation type to go for depends on several factors. First, you need to consider your research stage. If you have substantial results, interpretations, and conclusions, we recommend an oral presentation. However, if your research is ongoing and results are more preliminary, a poster may be more appropriate. Second, if you find public speaking extremely daunting, it may be worth presenting a poster to gain vital presentation experience in a less intensive atmosphere than giving a talk. However, you should remember that most people are terrified of public speaking, but this is an essential aspect of academia (and many other careers), and it gets easier the more you do it. Therefore, it's good to push out of your comfort zone and get as much presentation experience as possible. Third, if you are attending your first conference and are not sure of oral presentation styles and formats, then presenting a poster may be beneficial to allow you to gain more knowledge by attending other people's talks. While both poster and oral presentations make great additions to your CV, giving a talk is perceived more favorably. Furthermore, there are generally more poster spaces than oral presentation slots, so if your oral abstract is rejected, you can be asked to prepare a poster instead. Therefore, it's worth submitting an oral abstract if you are unsure which presentation type to apply for.

Most conferences have specific guidelines and requirements for abstracts, as they will be printed in the conference handbook. Therefore, it's vital to closely follow these guidelines and ensure your abstract adheres to the specified format, word limit, and any other requirements. Many conferences have an abstract template document you can use. The most important aspect of preparing a conference abstract is clearly and concisely summarizing your research within the limited word count (typically 150–500 words). You should generally follow the same structure as a research article (Introduction, Methods, Results and Discussions, Conclusions), but each section may only consist of a few sentences. Therefore, it's crucial to ensure your audience understands: (1) What's the problem, (2) Why it's important, (3) What you did to solve the problem, and (4) Why your results are significant. If you can clearly relay these points, your abstract will be in good shape. You should also tailor your abstract to the conference audience. For instance, if the audience is very broad, adapt your abstract by introducing your research field more and providing a less detailed analysis of your results. Conversely, if the audience is more specialized, your introduction can be more concise, and you can use more words to discuss your technical findings.

If the conference guidelines allow it, including a figure in your abstract can be an excellent way to present your results in a clear and visually appealing manner. However, be selective with your figure and only use one that clearly shows important results. Furthermore, ensure the details in the figure are big enough to read when the abstract is printed in the handbook and check whether the handbook will be printed in black and white or in full color. Depending on the conference guidelines, you may also want to include key references in the abstract. However, these should be limited to maximum of three and be highly relevant to your work. You should also create an accurate and compelling abstract title that grabs the reader's attention. Remember, your abstract may be one of several hundred at the conference, and attendees may only skim through the handbook; therefore, a catchy title and visually appealing figure can help interest readers. Finally, your abstract should be carefully proofread to ensure it's clear, has a good flow, and suitably communicates your research. If the organizers think the abstract is sloppy or unclear, they are far less likely to accept it.

## *Preparing a Conference Paper*

Some conferences also allow attendees to submit a conference paper, typically accompanied by an oral presentation. Conference papers generally have the same format as journal articles (see Chap. 6), but they are shorter and focus more on high-lighting key points and insights. Unlike journal articles, they are typically used to provide researchers with a platform to share their preliminary results, innovative ideas, and ongoing research rather than comprehensive and detailed studies. Furthermore, they usually have a less rigorous peer-reviewing process than journal articles, meaning they are perceived less favorably. Conference papers are published in the corresponding conference proceedings (handbook and online) and may be openly available on the conference website or an online repository. Due to this method of dissemination, they will generally receive less attention and fewer citations than journal articles. Although conference papers will not benefit your academic reputation and career in the same way as journal articles, they can still be a good way to disseminate your research, particularly preliminary results. Furthermore, these preliminary results may later develop into a full journal article. If you are interested in this, check the conference website to see if they accept papers. Furthermore, there may be a template document you can use to ensure your paper has the correct format.

More detailed advice on preparing oral and poster presentations for scientific conferences is given in the following chapter.

## *Creating a Plan*

Before attending a conference, it is important to prepare a plan. We suggest you print or download the conference proceedings and highlight which talks and sessions you would like to attend. If you are attending a large conference, several sessions may run concurrently. Therefore, it is important to identify topics and speakers relevant to your research field, as you won't be able to attend everything. Although a priority should be given to your own research area, it is often advisable to attend presentations outside of your research field that you may have a general interest in. Attending talks in different research areas can sometimes stimulate new ideas or lead to new collaborations. You should also make a note of the times and places of any poster sessions, workshops, training opportunities, conference dinners, and other events you are interested in attending. This will mean you end up with an approximate timetable you can use to guide your conference experience and ensure you get the most out of it. You can also identify any gaps in your timetable that may be good for sightseeing, catching up on emails, or networking with colleagues. The general dress code at conferences is usually smart-casual, but this is more of an unwritten rule and is not enforced. However, adhering to this dress code is usually a good idea. We also suggest creating a short list of key people you would like to connect with during the conference to help guide your networking (more on this later).

## During the Conference

The main thing you will do at a conference is attend presentations. You can use your plan to dictate which ones to go to. It can be very beneficial to ask questions at presentation sessions, although we appreciate this can be daunting at first. Furthermore, keeping detailed notes of talks can help you concentrate and ensure you have a record of any useful information presented. In addition to presentations, you will also spend your time doing several other activities. For instance, graduate students should prioritize poster sessions. This is because they are a great way to see what other early career researchers are working on in the field, generate new ideas, and make new connections. A bonus is that refreshments are often served. You may also attend training events or workshops to learn about various topics, including analysis methods, experimental techniques, and writing high-impact papers. If the conference is large, it may also have an exhibition hall, which is a designated space where scientific organizations, publishers, equipment manufacturers, research institutions, and other relevant entities set up booths to showcase their contributions to the scientific community. You can browse the exhibition hall to explore the latest scientific advancements, discover new tools and technologies, and connect with fellow researchers and professionals.

There are also many social/networking events you can participate in during conferences. For instance, there will generally be a welcome event on the first evening with refreshments and a short speech, in addition to several tea/coffee breaks and lunch each day. The cost of these is usually included in your registration fee, so it's worth making the most of them. Moreover, they are an excellent place to catch up with people and make new connections. Remember, if you have any specific dietary requirements, there is usually a place to mention this during registration. The organizers may also plan social events, such as sightseeing or evening activities, which can be an excellent way to make the most of your trip and broaden your social and professional networks. Many conferences also hold a conference dinner, which may or may not be included in your registration fee. The conference dinner is a formal or semiformal social event that takes place on an evening outside the conference venue (usually a hotel, restaurant, or event space). It is an excellent opportunity to build connections and to get to know people in your research area. They are often accompanied by speeches and can have lots of free alcohol—so be careful not to overdo it! After the conference, we recommend preparing a summary sheet that outlines any key things you have learned, ideas generated, and connections made, as this can be an excellent way to summarize your experience for your supervisor and lab mates, as well as for your own future reference.

## Networking

Networking is the process of building and maintaining professional relationships with people who share your research interests, career goals, or scientific pursuits. It can be crucial to the academic and professional development of graduate students, as it can facilitate idea generation, feedback on your research, exposure to new perspectives, fruitful collaborations, and career opportunities. However, despite its importance, PhD students typically receive little or no training in networking, and thus, it can feel uncomfortable, confusing, and daunting. Therefore, this section provides a comprehensive overview of how to maximize your networking opportunities during your PhD and beyond (Fig. 8.2).

### *Networking Environments*

During your PhD, you will have many opportunities to network with various stakeholders in and outside your scientific field. It is important to recognize these opportunities and understand which networking techniques are most

**Fig. 8.2**  It is important to develop good networking skills to increase your visibility and find new opportunities within your field

appropriate to implement. Some examples of typical networking environments are given below.

- *Conferences:* These events provide scientists with an environment for fruitful networking with a large and diverse range of stakeholders from their research community, including academics (from students to professors), industry members, and individuals from different sectors (*e.g.,* clinicians, policymakers, and government employees). Despite the diversity of attendees, they all share common interests, which creates extensive opportunities for insightful and valuable discussion. Therefore, conferences present an important environment for scientists to expand their professional network, develop research ideas, and enhance their career prospects. Networking at conferences generally occurs at lunch and refreshment breaks, poster sessions, and conference dinners.
- *Workshops and seminars:* Attending workshops or seminars, whether inside or outside your own institution, can provide an excellent opportunity to network with individuals working in a similar research field. These events are typically much smaller than conferences, which limits the audience scope but can make it easier to approach and start conversations with other attendees.
- *Departmental events:* Many academic departments hold events, such as social gatherings, guest lectures, and staff meetings. These provide excellent opportunities for networking with various researchers within your department, which can be extremely helpful in terms of fostering collaborations, accessing equipment, research support, and career opportunities. Therefore, PhD students should try to attend these types of events if possible.
- *Lab meetings:* Many professors hold regular meetings for their laboratory members, which may include students, postdocs, and visiting scientists. Consequently, they can be an excellent environment to foster close relationships with your lab mates, expand your research knowledge, and explore the potential for collaborative research.
- *Day-to-day work:* During a typical working day, you may interact with many individuals, from students to technicians to faculty members. Again, this can present excellent opportunities to strengthen your network, gain experience in other people's research areas, and generate valuable ideas. Moreover, it can help you gain access to equipment and potentially secure your next academic position. Therefore, it is crucial to always be forthcoming, proactive, and approachable within your workplace.
- *Research organizations and institutes:* Many research organizations and institutions host events, workshops, and seminars that may take place locally. These events generally attract numerous researchers who may work in somewhat different areas but are linked through the organization or institution (*e.g.,* Royal Society of Chemistry or American Chemical Society). This creates ample networking opportunities with individuals based locally that are inside or outside your research field.

- *Outreach events:* Attending or organizing science engagement events can be an excellent way to connect with members of the general public interested in science or with other researchers with a passion for science communication.
- *Online:* Networking online is becoming increasingly crucial for all researchers, as it offers a convenient and expansive way to connect with various stakeholders across the globe. Engaging on platforms such as ResearchGate, LinkedIn, and X allows scientists to share their research, insights, and ideas. Moreover, it provides an excellent environment to gather feedback on work, facilitate useful discussions that transcend any geographical limitations, and learn about cutting-edge research within your field. Online networking provides a great way to find career opportunities or establish new collaborations without costly and time-consuming travel to different in-person events.

## *How to Network*

Below are some tips you can use to improve your networking skills. They can be incorporated into any networking environment and will help to create mutually beneficial connections with various stakeholders. However, remember that networking is a skill developed over time, so don't put too much pressure on yourself to be fantastic at it immediately. Instead, you should try to gradually improve your networking skills by incorporating some of the tips below into your behavior. Additionally, remember that almost everyone finds networking awkward, anxiety-provoking, and uncomfortable at times, so having these feelings is completely normal. As with most things, the more you do it, the easier it gets.

### See It as Work

Before you do any networking, it's important to understand exactly what it is. There is a large difference between networking and making friends, which is something many people fail to understand. Networking is about making professional connections and building rapport with individuals in a limited timeframe. Yes, you may become friends with people in your network and you should always be friendly, but networking itself should be professional with a clear purpose. Furthermore, understanding that networking is a professional activity and part of one's job role can help alleviate anxiety and awkwardness.

### Be Prepared

Before attending events, you should have a clear plan of who you want to connect with and what you want to achieve. For example, are you looking for new collaborators, seeking job opportunities, or trying to keep up to date with the latest research in your field? Creating specific objectives will help provide direction and structure to your networking. If you are attending an event where the attendee list is available beforehand (*e.g.,* a conference), you should go through this list and identify individuals you would like to connect with. Furthermore, do some research on these individuals and their work. For instance, being complimentary about someone's recent paper and asking them a couple of relevant questions shows that you value them and have a legitimate interest in their work. However, you should also be open to unexpected connections, as these can be really valuable.

## Right Time, Right Place

To maximize networking opportunities, it is important to make yourself available to connect with new people. For example, at conferences, you should attend planned events such as coffee breaks, formal dinners, drinks receptions, lunches, and poster sessions. These are excellent environments to build connections, as networking is expected. Furthermore, approaching people at their posters or after they have given a talk provides a great foundation for networking, as you can initially discuss their poster or talk. However, you should also focus on networking in your day-to-day working life. For example, eating lunch in a communal area or attending after-work drinks can create opportunities to build valuable connections with different members of your institution.

## The Elevator Pitch

You should prepare a brief, engaging introduction about yourself and your research. This will help to quickly create meaningful connections with people and establish common interests (Fig. 8.3). You can tailor your pitch to the event you attend or the individual you speak to. For example, you can add more technical detail if you are at a conference focused on your research area. However, if you are talking to someone from outside your field, you should minimize technical language to ensure you are easily understood. The pitch should last about 30 seconds and follow a general format of introduction, your role and affiliation, your research area, your research impact or benefits, and a closing statement with a question or call to action. For example, my elevator pitch (JM) may be: "Hi there, it's a pleasure to meet you. My name is Jake McClements, and I am an Academic Track Fellow at Newcastle University. My research is focused on developing next-generation wearable sensors to improve human health and well-being by tracking certain diseases. I would love to discuss this further and explore any potential collaborations." If you already know of the person, then you can make your pitch more personable. Furthermore, mention if you have a previous connection with them, such as a mutual friend or colleague. Although elevator pitches can seem awkward initially, they are an excellent way to quickly and efficiently connect with people. It is advisable to practice a pitch before attending events but be careful not to come across as robotic when saying it.

**Fig. 8.3** Having a pre-prepared elevator pitch is a good way of starting a conversation with someone at a conference

## Introductions

A great way of facilitating networking is through introductions by a mutual connection. For example, your supervisor may introduce you to several key people in your field at a conference, which can help alleviate the anxiety of walking up to someone alone. Furthermore, an introduction from a mutual contact helps form connections and generate talking points. However, don't ask the same person to perform too many introductions, as this may become awkward for them. You can also 'share networks' with someone, where you each introduce them to some different people you know to help facilitate networking. If you don't have anyone to introduce you, this isn't a problem. All you have to do is approach someone, smile, and politely introduce yourself.

## Names

If you're networking at a conference, ensure your nametag is visible. Furthermore, read other people's nametags and use their names in your conversation. If you are in a more informal setting without nametags, introduce yourself and ask for someone's name. Using names in conversations is an excellent way to be remembered and make more personable connections.

## Be Personable and Professional

One of the most important factors in successful networking is how you carry yourself when talking to people. It is important to be polite, professional, and friendly. This will ensure you leave a good impression and do not offend or upset anyone. Remember, you do not want to be remembered for the wrong reasons or build a bad reputation in your field. Therefore, if you are not getting along with someone, it's far better to politely remove yourself from the conversation than create conflict with them. Likewise, don't be too intense or overbearing with people; sometimes, it's best to keep conversations relatively brief rather than talking at someone for a long time when they may not be that interested or want to move on. If you are networking at a conference, alcohol may be involved (*e.g.,* drinks reception, conference dinner). If this is the case, try not to overdo it, as it's difficult to remain professional if you're drunk! Remember, networking and socializing with friends are two different things.

## Confidence

It is easy to say "be confident and outgoing"; however, this can be exceedingly difficult for some people, such as introverts or those not speaking their mother tongue. Therefore, if you are not confident in these situations, start very small and gradually build your confidence. For example, if you're at a conference, go to a poster session, introduce yourself to a couple of presenters, and ask some small questions about their work. You do not have to be networking constantly; you could just try making one or two connections each day.

## It's Give and Take

Remember, networking is not all about what someone can do for you; it's also about what you can do for them. This is something that many people get wrong. You also have a lot to offer, which could be valuable to people and create strong connections. For example, you may have expertise in a specific experimental technique (*e.g.,* atomic force microscopy) and could offer to perform some measurements for someone. This may lead to coauthored journal articles and a strong, ongoing collaboration. Before networking, think about your value and what you can offer people. However, don't overpromise; it's important that you are always realistic in what you offer.

**Choose the Right Person**

If you are a graduate student, it can feel like experienced professors are not always interested in networking with you. They often know many people at the event, are approached by many graduate students/postdocs, and have limited free time. It is therefore advisable to carefully select who you approach. For example, you may want to connect with early career researchers, such as Assistant Professors, who are beginning to build a reputation in their field but are not yet firmly established. Connections with these individuals can be more meaningful, as they are often actively seeking new lab members or collaborators. Furthermore, there is a greater chance that you could offer them a valuable service or expertise. Therefore, numerous forgettable conversations with esteemed professors may be less valuable than one or two meaningful connections with early career researchers. If you are networking with esteemed professors in your field, then it's best to research their work and have some questions prepared.

**Ask Questions and Listen**

When networking, don't just talk about yourself. Show genuine interest in other people's work by asking questions and listening to their answers. This helps to build meaningful connections with people and leave a good impression. Furthermore, it can alleviate awkwardness, as people generally like to talk about themselves. Therefore, if you find the conversation stalled, ask someone a few questions to help move things along. However, you should not bombard people with questions. Instead, you should carefully listen to their answers and let the conversation evolve naturally.

**Follow-Up**

One of the most important aspects of networking is following up on connections you have made. This will help cement connections and significantly increase the probability of forming a mutually beneficial relationship. Following up simply means contacting someone after initially meeting them. You can initiate this by asking for someone's email, adding them on LinkedIn, or giving/receiving a business card when networking. You can then politely contact them by referencing your connection, expressing gratitude for their time and insightful thoughts on your work, providing value by sharing relevant resources or research ideas, and proposing a next step, such as setting up an online meeting to discuss a potential collaboration. You don't have to do this with everyone you have networked with, only when there is potential for a meaningful ongoing connection. It is easy to forget every connection you have made, so being proactive and following up on potentially valuable connections is highly recommended.

## Have an Exit Strategy

Typically, you will have limited time for networking at an event. Therefore, preparing an exit strategy to leave a conversation if it is too long and you want to network with others is a good idea. This can be difficult, and you don't want to be rude or awkward, so having an established exit strategy can be very useful. The website *Science of People* has developed an Exit Formula that can be used to great effect:

$$Genuine\ Compliment + Followup\ Item + Handshake\ (or\ gesture)$$
$$= Lasting\ Impression$$

When you want to leave a conversation but cannot find a natural end, begin by professionally complimenting the person you are networking with. For example, "It's been a pleasure speaking with you," "It's been really interesting hearing about your work," or "Thanks again for an insightful talk." You can then discuss a followup item, such as "I will find you on LinkedIn" or "Hope to see you at the conference dinner." If it's a potentially meaningful connection, ask them for an email address or business card and say you will contact them for further discussions. Finally, a handshake is a clear cue that the conversation is over, and you can politely exit by saying goodbye or something like "Good luck with your talk later." If a handshake is inappropriate or you do not feel comfortable doing this, a similar gesture is fine, such as nodding/bowing your head. This method will ensure that you maximize your networking time and leave positive, lasting impressions.

## Social Media

Not all networking occurs in person, and social media can be an excellent way to make new connections globally. For example, you may use a conference hashtag and post about your experience to help connect with other attendees. Likewise, posting about your recent papers or achievements can help you connect with others in your field. If you believe a connection on social media is valuable, you can politely message someone and ask for their email address to facilitate a more in-depth discussion.

Networking is something that many people find awkward and anxiety-provoking at first, but it's a skill, and the more you practice, the better you get at it. Therefore, try to be conscious of these tips when networking and start small to gradually build your confidence and skills. Remember, most people also feel awkward in these situations and often appreciate someone approaching them to create a connection. Ultimately, you have little to lose by networking but could potentially form new collaborations, find future job opportunities, or receive valuable research ideas. Consequently, we strongly recommend that you go out of your comfort zone and maximize your networking opportunities throughout your graduate studies.

## Concluding Remarks

Conferences are an important part of academic life, as they provide a platform to showcase your research, gain presentation experience, receive valuable feedback, establish connections, stay up to date on the latest developments in your field, and gain life experience by traveling to new places. However, it is important to choose the right conference for you and maximize your experience there by preparing a poster or oral presentation, as well as attending relevant talks, poster sessions, and conference dinners. Networking is also a vital aspect of academic life, as building connections with a diverse range of stakeholders will help you form new collaborations, find job opportunities, and gather useful insight into your work. Networking can be difficult at first, but starting small and practicing regularly will help you improve your skills and gain confidence.

**Key Points**
- Attending scientific conferences is important for learning about the latest developments in your field, as well as for networking with other scientists.
- Scientific conferences are often expensive to attend and so it is important to carefully select the most appropriate one.
- It is often useful to create a plan before you go to a conference so you know which talks and posters to attend, and who you would like to connect with.
- Presenting your work as a poster or talk at a scientific conference increases your visibility and impact, as well as providing constructive feedback on your work.
- Networking with other scientists is critical as it can lead to new collaborations or job opportunities.
- Networking can be challenging but it is a skill that can be learned and developed over time.

## Resources

Science of People. How to network at a conference: 10 ways to make contacts like a pro. (2019). https://www.scienceofpeople.com/conference-networking/

# Chapter 9
# Preparing Talks and Posters: Effectively Communicating Your Research Findings

## The Importance of Effective Communication

Developing effective communication skills is critical for any scientist. There is no point in doing good research if nobody ever learns about it. The better a communicator you are, the more impact your research will have. There are many situations

© The Author(s), under exclusive license to Springer Nature Switzerland AG 2024          193
D. J. McClements et al., *How to be a Successful Scientist*,
https://doi.org/10.1007/978-3-031-51402-9_9

during your graduate studies where you will require good oral communication skills:

- *Laboratory meetings*: Many professors require their students to periodically give presentations about their research findings in lab meetings. This is an excellent opportunity for students to develop and refine their oral presentation skills. The professor and other lab members can provide constructive feedback on the design of your presentation and your oration skills.
- *Graduate seminars*: Many graduate programs require students to give one or more oral presentations as part of their degree. The topics of these presentations may be based on the student's research area or something different. Again, this is a good opportunity for students to obtain feedback from their peers and professors to improve their presentation skills.
- *Scientific conferences*: During and after your graduate studies, you may be expected to present your research findings at national or international scientific conferences. Many experts in the field often attend these conferences, so it is important to present your research in the most effective and engaging manner possible.
- *Oral thesis presentations*: As part of your graduate studies, you may have to present your research findings to your graduate committee. In most programs, this is performed at the end of the degree as part of the final exam, but it may also be done part way through the degree, *e.g.,* for a prospectus exam.
- *Workplace presentations*: If you are employed by a company or the government after graduating, you may have to give oral presentations as part of your work. You want to give informative, clear, and engaging presentations, as this will improve your standing in the workplace.
- *Online presentations*: You may be asked to give scientific presentations that are accessed online. For example, as part of your website or when you are developing an online course.

These are only a few examples of situations where you may have to give an oral presentation. Giving a talk that leaves a strong positive impression on your listeners can greatly benefit your reputation and career prospects. For instance, if you make a good impression at a conference, you may be able to find collaborators or people who are willing to give you a job in the future.

## Oral Presentations

The best way to develop good oral communication skills is by critically observing other people's talks, as well as giving your own. When attending a scientific talk, you should think critically about what you like and dislike about the speaker's presentation manner and the design of their presentation. However, the best way to learn how to give good oral presentations is to do them!

## Types of Scientific Talks

In general, there are two major types of scientific talk: research talks and overview talks. These two presentation types generally correspond to original research articles and review articles, respectively:

- *Research talks*: These talks usually focus on a particular piece of research that the speaker has carried out. They roughly correspond to the results that might appear in an original research article. For example, they may describe the results of a study on preparing and characterizing a new polymer.
- *Overview talks*: These talks typically provide an overview of a particular area. They roughly correspond to the content of a review article. For instance, they may describe the preparation, characterization, and application of nanoparticle-based delivery systems in the pharmaceutical industry. Therefore, they may rely on research performed by the speaker and research published by others.

The aims and design of research and overview talks are different and should be carefully considered during their preparation.

## Preparing Scientific Talks

Good preparation is critical for creating a successful scientific presentation. You should give yourself sufficient time to prepare the talk – never wait until the last minute, as this will mean you do not have time to create, learn, and refine your presentation. Typically, you want to leave at least two weeks to prepare, and sometimes considerably longer. There are several things you should find out before preparing your presentation, as they will help you design your talk: When and where will your talk be given? How long will it be? What is the nature of the audience?

Some of the most important steps in preparing an effective scientific presentation are given here:

- *Topic selection*: You should select an appropriate topic to discuss, which will depend on whether this is a research or overview talk.
- *Title selection*: Once you have selected a suitable topic, write a clear and compelling title. Developing a good title helps you to identify the most important content and structure for your talk.
- *Synopsis*: It is advisable to summarize the goal of your talk in two or three sentences. What is the key message you want the audience to take away? This will help you to focus on the most important aspects. For instance, the synopsis of a talk on creating a new polymer may be "This presentation aims to explain how a new polymer was designed, fabricated, and characterized. It also aims to highlight the advantages of this new polymer over existing ones."
- *Topic familiarization*: Once you have selected a topic to present and have a good idea of the aims of your talk, you should then thoroughly familiarize yourself

with the topic. If it is a research presentation, then this should be relatively straightforward. However, if it's an overview presentation, you may have to conduct an extensive literature search to identify all the relevant material you want to present.

- *Talk Structure*: It's good practice to develop a rough draft of the overall structure of your talk before going into the details. What main points do you want to include in the introduction, main body, and conclusion sections? It is often a good idea to create a list of bullet points for each section, as this will help to structure your talk.
- *Slide preparation*: Once you have an overall structure, you can start creating each slide you intend to present. This should usually be done using a high-quality presentation software program (such as Microsoft PowerPoint).
- *Practice:* It is important to practice a scientific talk several times before you give it, especially if you are not used to giving them and/or it is going to be presented at a major meeting or conference. Initially, this could be done on your own, but it is preferable to do it in front of your colleagues, such as other lab members or your advisor. You may also want to practice it in front of your family and friends, as they can give you some valuable insights into its clarity and flow. The feedback provided by others can help you refine and improve your presentation.

## Make It Interesting!

The main purpose of your presentation is to convey important scientific information to the audience. You want to tell the audience what you did, what you found, and why it is important. However, it is also important to make your talk engaging and memorable, as this will enhance the impact of your research, as well as your visibility and reputation in your field of expertise. For instance, your talk may be on the last day of a large conference where most of the audience has already sat through multiple talks. Therefore, it is important to ensure that your talk stands out and gives the audience something to remember about you and your work.

We introduced Randy Olsen in an earlier chapter (Chap. 6). He was the University of Maine professor who studied Marine Biology but gave it up to become a film director in Hollywood. He has written several books encouraging scientists to improve their communication skills so they can engage the public and other scientists better, thereby increasing the impact of their work. In his book *Don't be Such a Scientist,* Olson criticizes many scientists for giving talks packed with dense information and presented in a way that the audience finds overwhelming, uninteresting, and tiresome. As a result, the speaker rapidly loses the audience's attention, so they learn little from the talk. Olson argues that it is far better to provide much less information but to present it in a visually and orally compelling fashion that keeps the audience's attention. He emphasizes the importance of having a narrative that runs through the talk and telling compelling stories that the audience finds interesting. The audience wants to know why the information being presented is important to them. Ideally, you should plan your talk to contain at least one "hook" that you think the audience will remember. If they remember the hook, then they are more likely to remember you and your talk. This hook could be an impactful or compelling image, video, fact, or research finding.

In his later book, *Houston, We Have a Narrative*, Olson introduces a structure for making scientific talks (and scientific manuscripts) more interesting that he calls the ABT method, which stands for "And…But…Therefore." He contrasts this with the AAA method, which stands for "And…And…And." Olson argues that scientific talks fail for two main reasons: they are too boring or too confusing. In an AAA talk, the speaker provides too much information without any context or story: this happened AND then this happened AND then this happened AND then… The audience just hears lots of facts, which can be overwhelming, boring, and confusing. As a result, they switch off, and your presentation has little impact. In contrast, in an ABT talk, the speaker structures their presentation to increase the audience's interest and attention:

- **AND**: The speaker begins by introducing the context of the study by providing a series of statements: "The global population is growing AND people are becoming wealthier AND eating more animal-based foods…"
- **BUT**: The speaker then introduces a problem or question that needs to be addressed, which raises the tension and the audience's interest: "BUT, this is leading to an increase in greenhouse gas emissions, pollution, and biodiversity loss."
- **THEREFORE**: The speaker then introduces how the problem or question may be addressed and the tension resolved: "THEREFORE, we need to reduce the consumption of animal foods by creating a new generation of plant-based foods."

This approach is similar to how a writer might structure a story to make it more compelling: exposition, conflict, and resolution. First, the writer sets the story's scene by providing information about the time, place, and characters involved. Then, they introduce some conflict that creates tension, which increases the reader's interest. Finally, they resolve the conflict by addressing the source of tension. You should therefore design your scientific presentations with the ABT model in mind, as this will make them more interesting and engaging. Chris Anderson, the head of the TED organization, makes a similar argument in his book *TED TALKS: The Official TED Guide to Public Speaking*. The ABT approach is highly impactful in the introduction to your talk, but it can also be used for each of the themes that you address in the main body (see later).

## Know Your Audience

Before preparing your scientific presentation, knowing your audience is important, as this impacts how you design and present your talk. In particular, it is essential to know the audience's level of knowledge and interest in your research. In some scientific meetings, the audience may be experts in your area. Therefore, you may want to give introductory remarks that are more specific rather than more general. Moreover, you may not need to describe your methods in detail because everybody is already familiar with them. As an example, if you were giving a talk on plant-based meat substitutes at a conference on food sustainability, you do not need to spend much time talking about the negative impacts of the food supply chain on the environment because everybody already knows this. Furthermore, this information has likely been repeated numerous times in the introductions to previous talks, which can be boring to the audience. In contrast, at a scientific meeting where the members of the audience are not experts in your area, you may want to give introductory remarks that highlight the general relevance of your research and provide more background information about the methods you used because the audience may be unfamiliar with them. For example, if you were talking about plant-based meat substitutes at a polymer science meeting, where few people have knowledge about foods, then you may talk about how livestock production is contributing to greenhouse gas production, biodiversity loss, and pollution at the beginning of your talk to highlight the importance of your research.

You do not want to give a highly technical talk with many details to a general audience, as they will not understand and may not be interested. Conversely, you do not want to spend time telling people what they already know. If you are talking to an audience who already knows your area well, then focus on the new things you have found rather than spending a long time on general material. Pitching presentations to an audience at an appropriate level is a critical skill of being a successful scientist, which is also important in numerous other areas, including collaboration, teaching, supervision, and oral exams.

## Audience Attention

The audience's attention is usually greatest at the beginning and end of a talk (Fig. 9.1). Consequently, it's crucial to design your talk to make the introduction and conclusion sections highly impactful and engaging. An audience often loses attention in the middle of your talk. This can be overcome by structuring your presentation to contain several themes with their own mini-introductions and mini-conclusions that help maintain the audience's attention. As discussed earlier, you can use the ABT structure for each theme – set the scene, introduce a conflict, and then show how it is resolved.

## Structure of an Oral Presentation

This section will provide advice on how you can structure a themed presentation to engage the audience. Many students design their scientific presentations in a format similar to a scientific manuscript. For example, for a research talk, they may include

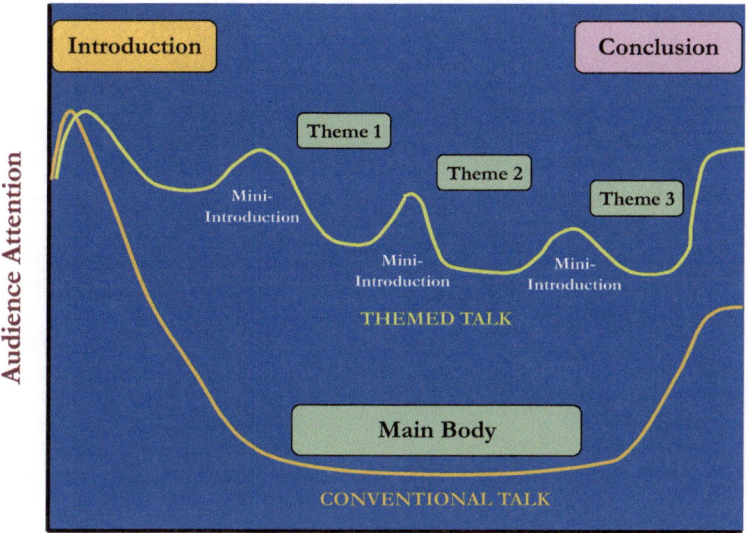

**Fig. 9.1** In a conventional talk, the attention of an audience is usually greatest at the beginning (introduction) and end (conclusions) and sags in the middle (main body). It is therefore a good idea to do a themed talk that breaks the main body into a few themes with their own mini-introductions and mini-conclusions to keep the audience engaged

Introduction, Materials and Methods, Results and Discussion, and Conclusions sections in their presentation (Fig. 9.2). This is rarely the best way to give a compelling oral presentation that engages the audience. If you use this format, your presentation will not flow smoothly and may not tell an engaging story. As mentioned earlier, it is better to use a themed talk that is more compelling to listeners but still conveys essential information. Usually, people are not interested in all the details of your research; instead, they are more interested in your big-picture findings and their implications. If the audience can remember one or two things from your talk, then it has been successful. In the remainder of this section, we discuss the various parts of a good scientific presentation.

## *Title*

The title of your presentation is important for several reasons. Most importantly, it provides people with information about the content of your talk. If you are presenting at a scientific conference, a good title will help to attract people to your talk, thereby increasing the impact of your research findings. The title of the presentation will also appear in any program or website associated with the meeting, as well as in your *curriculum vitae*. It is therefore important to choose an informative, concise,

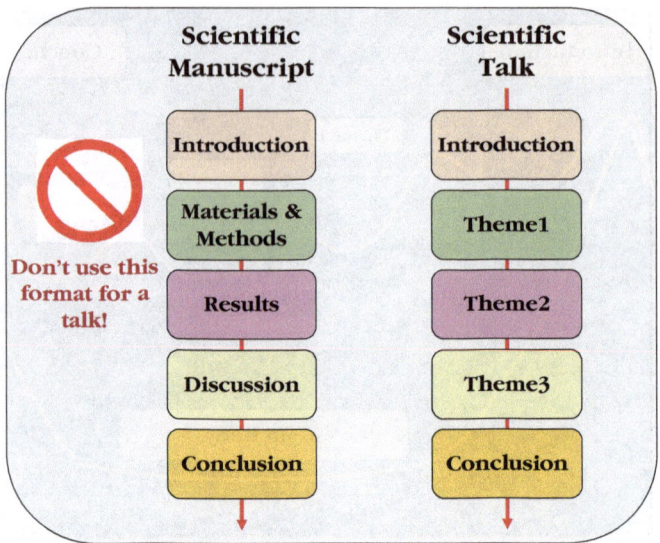

**Fig. 9.2** A scientific talk should **not** be structured the same way as a manuscript. It is much more engaging for the audience if you tell a compelling story about your research findings. This can often be achieved by breaking the main body of your talk into different themes

and compelling title. It's usually a good idea to keep your title quite general to attract a broader audience, but this depends somewhat on the nature of the audience you are addressing. The more specific the expertise of your audience, the more specific your title should be. For instance, if you were talking at a biopolymer conference, you may use the title "Utilization of gellan gum and potato protein to form composite hydrogel matrices suitable for application in meat analogs," but if you were giving the same talk at a general food science conference, you might use the title "Creation of plant-based meat analogs from food proteins and polysaccharides."

## *Introductory Remarks*

The introductory remarks are one of the most important parts of your scientific presentation, as this is where the audience has the most attention (Fig. 9.1). Your aim should therefore be to get the audience excited (or at least interested) in the information you are about to present. After hearing your introduction, the audience should know what you will discuss and why it's important. Usually, the session's moderator will introduce you and say where you are from, so you don't need to repeat this information. You may start by saying something like "Good morning/afternoon/evening everybody. I would like to begin by thanking the conference organizers for allowing me to speak. I am looking forward to presenting some of my recent research findings." You then want to introduce the topic you will talk about. You do not want to simply read the title of your talk from the screen - the audience can already see this, and the moderator may have already mentioned it. Instead, you want to briefly paraphrase the title in a succinct and engaging fashion.

There are several elements you should include in your introduction section, with the precise nature of these elements depending on the topic of your talk and the members of your audience (see earlier):

- *Relevance*: You should highlight the importance of the general topic you will discuss. For instance, if you were giving a talk on creating plant-based meat substitutes, you may include one or more slides describing the negative impacts of livestock production on greenhouse gas emissions, pollution, and biodiversity loss. This slide lets the audience know why the problem you are discussing is important and hopefully gets them interested in the topic. If you are talking to a general audience, then this may include several slides, but if you are talking to a specific audience who is already familiar with this information, then this could be a single slide. As mentioned earlier, you should not waste time telling people what they already know.
- *Approach*: You should then highlight the approach you will take to address the problem you have outlined. For instance, in the example just given, you may include one or more slides introducing the concept of soft matter physics and how it can be used to create meat-like substances from plant-derived ingredients.
- *Outline*: Some speakers like to include a list of things they will cover in their presentation immediately after the introduction, as this helps the audience know what to expect. However, doing this is often redundant because this information will be presented during the talk. Instead, it may be better to just give an oral summary. For instance, "In this talk, I want to show you how soft matter physics can be used to produce meat substitutes from plant-based ingredients." Alternatively, you may want to provide a more visual representation of what will be covered in the talk rather than using only text (Fig. 9.3).

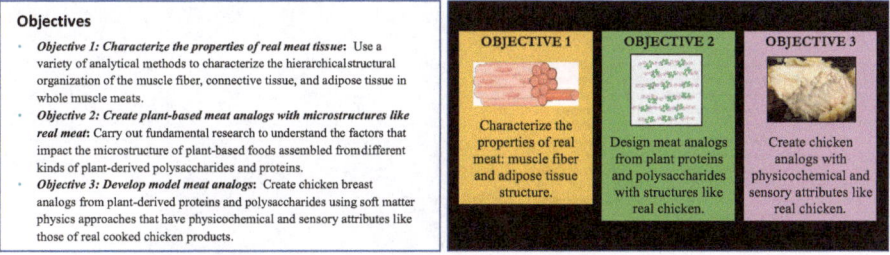

**Fig. 9.3** Near the start of their talk, some speakers list the topics to be covered in the presentation. Rather than using all text to do this, it is usually better to use diagrams. The slide on the left contains more information but is difficult to read and visually unengaging. The slide on the right is much more engaging

## *Main Body*

The main body of the presentation is where you present your research findings. Typically, it is not advisable to discuss all the materials and methods together, followed by all the results and discussion, as you would do in a scientific manuscript. Instead, it's better to break the main body of your talk into several subsections based on different themes, with each theme having a mini-introduction and mini-conclusion, as this will make your presentation more engaging to the audience (Fig. 9.1). Each theme may include a mixture of introduction, materials, methods, results, discussion, and conclusions. The goal is to make the presentation visually engaging and interesting rather than providing numerous details that overwhelm the audience and lose their interest.

## *Concluding Remarks*

The concluding remarks are another critical part of your presentation. This is the last thing your audience will hear. Moreover, the audience's attention usually rises toward the end of a talk (Fig. 9.1). It is therefore important to leave a strong impression on the audience by having a compelling and engaging conclusion. This will increase the impact of your presentation and ensure the audience remembers who you are and what you talked about. Your concluding remarks should contain several elements:

- *Summary*: You should give a brief overview of what your study found, typically in the form of concise bullet points accompanied by engaging images (Fig. 9.4).
- *Implications:* You should highlight the importance of your research findings – what was novel, what was unexpected, and what are the implications to your field of study or society as a whole.
- *Future work:* You may highlight areas where further research is still needed.

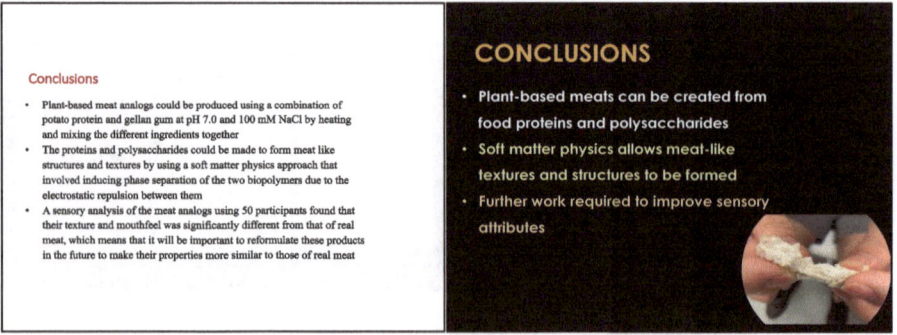

**Fig. 9.4** It is important to finish your presentation with an impactful conclusion. Rather than providing a detailed list of what you found that may be difficult to read (left), it is better to present a visually appealing concluding slide containing concise bullet points you can talk about (right)

Traditionally, speakers end their talks by including an acknowledgment to the other people who helped produce the information presented, such as professors, students, postdocs, and other collaborators (often a laboratory group photo is shown), as well as organizations that provided the funding that supported the research, such as government, industrial, or nonprofit agencies (often their logos are shown). However, some speakers prefer to include their acknowledgments at the beginning of their presentations. If you are a professor, or your work is strongly multidisciplinary, then you may include information about the people who carried out the research on the slides where their work appears, *e.g.,* by including their name and image. In some disciplines, especially medicine, it is important to include a disclosure of potential conflicts of interest near the beginning of your presentation.

## The Importance of Images

Compelling images greatly increase audience engagement. These could be well-designed diagrams, figures, animations, or movie clips. As a rule of thumb, having at least one image per slide is preferable. A slide full of text is usually difficult to read in the time available and visually unstimulating. Rather than overloading the audience with text, it is usually better to include the same information concisely with one or more images (Fig. 9.5).

A part of the presentation where diagrams can be particularly useful is when discussing the various materials and methods used. Rather than writing a detailed list of all of these, it is more engaging to present them using diagrams (Fig. 9.6). If anybody needs more details about the materials and methods, they can ask you at the end of your talk or read any scientific manuscripts based on your presentation.

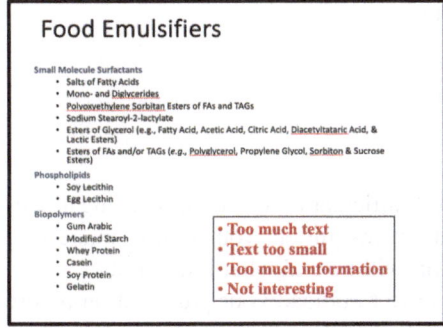 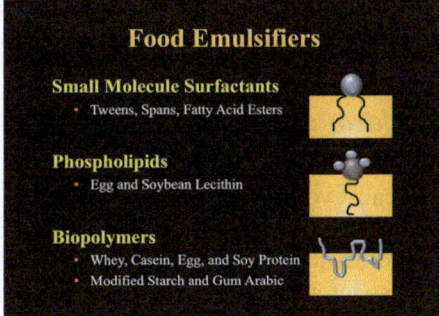

**Fig. 9.5** It is important not to include too much text on slides as this can be difficult to read. Presenting the same information in a more succinct form with images is typically much more engaging for the audience

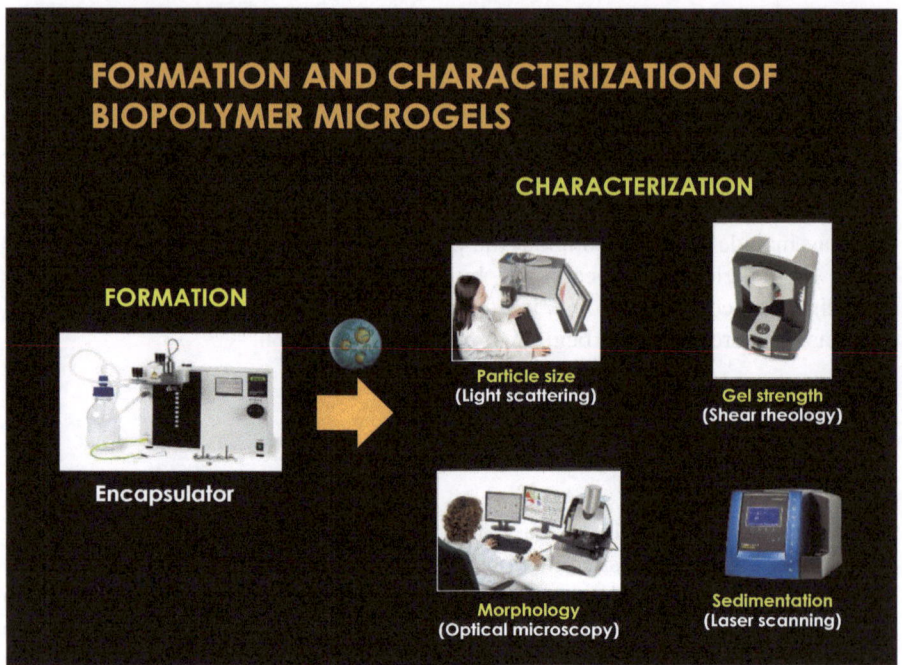

**Fig. 9.6** When discussing the materials and methods used in your research, it's better to use images rather than detailed text. Images kindly provided by Buchi, Malvern Panalytical, and Formulaction

## Practical Considerations

In this section, we provide some additional tips that can be useful for improving the quality and impact of your oral presentations.

### *Preparation*

Typically, your presentation will be given at a particular time and place, using audio-visual equipment supplied by the organizers or institution. You want to make sure you know when and where your presentation will be and what kind of audiovisual equipment will be available. You may also have to upload your presentation before giving your talk, so you should know how to do this. This may involve sending your presentation to someone by email, uploading it to a website before the meeting, or uploading it from a memory stick at the meeting. Always ensure you arrive well before you are expected to talk – you do not want to be late or rush into the room and have to talk immediately. You may also want to bring your own laser pointer in case the conference organizers do not provide one (or the batteries in theirs have run out).

## *Timing*

You should find out how long your talk is meant to be (*e.g.,* 20, 30, or 40 minutes) and how long you need to leave for questions. It's imperative that your talk fits into the allotted time. It's extremely bad practice to give a talk that is too long, as this affects the timing of all the other presentations and may mean that the session runs out of time. If timings are strict, then you may have to rush through the latter parts of your talk to finish it in the allotted time, which reduces your presentation's impact. Typically, you want your presentation to be 1 or 2 minutes shorter than the allotted time. It is therefore important to rehearse your talk to ensure it's the correct length. It is better to be too short than too long, as any extra time can be used for questions from the audience. As a general rule, you want about 1 to 2 slides per minute, depending on the complexity of the material covered. This will give you sufficient time to cover all the material on each slide without rushing. If the material to be covered is too complex for a single slide, then you should use more than one slide to clearly explain it.

## *Rehearse*

After you have prepared your presentation, it is important to rehearse it for several reasons. Ideally, you should give your talk without using any notes, so you need to remember the material to ensure your talk flows smoothly. As just mentioned, it's also important that your talk is the correct length, so practicing it can help you establish the running time. You can then add or remove slides if it's too short or too long. If possible, you should practice your talk in front of other people, such as your friends, lab mates, or professor. They can then advise you on what needs to be removed, added, or revised in your talk. They can also tell you if your presentation flows smoothly or the material is explained clearly enough. Often, when practicing alone, you may not realize things such as speaking too quickly or putting your hands in your pockets, which others can easily point out. Typically, the more nervous you are about speaking, the more practice you should do, as this will make it easier to remember your talk if you get stuck in some places.

## *Content Density*

Avoid overcrowding your slides with too much text or complex information. Aim for a balance between providing enough information and keeping it concise. Use bullet points, short phrases, and strong visuals to convey your main points effectively.

## *Font Size, Style, and Contrast*

Your slides should have good readability. It's therefore, important to use a relatively large font size so your slides can be read quickly and easily, as well as being visually engaging (Fig. 9.7). On Microsoft PowerPoint, the preferred font sizes are 28–40 for slide titles, 24–28 for headings, and 18–24 for regular text. However, smaller font sizes may be more suitable for some parts of the slide, such as diagrams or figures, as long as they can be clearly read.

Many different font styles are available in presentation software such as Microsoft PowerPoint. You should choose one that is easy to read, engaging, and sets the correct tone for your talk. Some font styles may be too fancy or modern looking for scientific presentations. Some commonly used font styles are Times New Roman, Helvetica, Verdana, Arial, and Calibri.

It is also important to ensure that there is a good color contrast in your slides. The colors of the text, images, and background should be carefully selected to ensure a contrast that is easy to read and visually attractive (Fig. 9.8).

## *Images*

As a rule, it is always a good idea to have at least one image per slide, as they can make it much easier for an audience to quickly grasp important concepts, as well as being more visually appealing and stimulating. However, images should be designed so they are clear, properly labeled, and easy to comprehend. Whenever possible, it is therefore a good idea to include images, rather than lots of text that an audience may find difficult to quickly read and comprehend (Figs. 9.4 and 9.5).

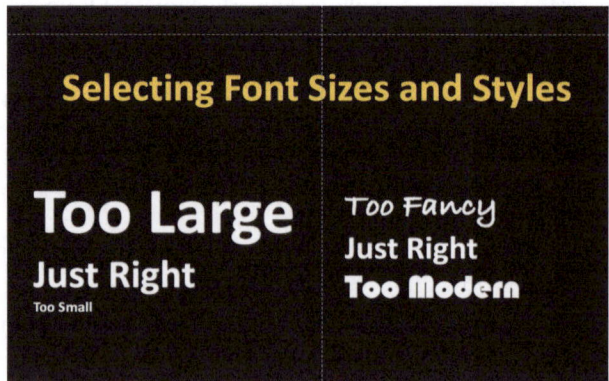

**Fig. 9.7** It is important to select a font size and style that is readable and visually engaging

**Fig. 9.8** It is important to select colors for text and images that give good contrast with the background used

## *Oration Style*

How you speak during your scientific presentation also has a major influence on audience interest, attention, and understanding. One of the most important things is to sound enthusiastic. If you are not excited about your research, you cannot expect the audience to be. You should not talk too loudly/softly or slowly/quickly. If you speak too quickly or softly, people may not understand you. If you speak too slowly, people may become bored and switch off. If you talk too loudly, then people may become irritated. It's important not to talk in a monotone manner – change your tone during your talk to engage the audience. It's also important to not just read the text on your slides – the audience can already see this information. It's better to paraphrase the material and present it more conversationally. You should also try not to move around too much when speaking, such as pacing backward and forward across the stage. However, it is desirable to use hand movements to emphasize important points and keep the audience interested, rather than keeping your hands in your pockets and staring at the screen.

If you have enough confidence, it's a good idea to interact with your audience. For instance, you could ask the audience a question and wait for their response or provide your own answer. For example, if you were giving a talk on creating more sustainable foods, you might ask "Has anybody in the audience eaten insect burgers? Raise your hand if you have." Or you may bring props to make your talk more interesting, such as an insect burger that you take a bite from during your presentation.

## Eye Contact

You should look at different people in the audience as you speak, as this will make them more engaged. Don't always look at the floor or the screen. However, you should not lock into a single person and talk at them for the whole presentation, as this may make them uncomfortable (or they may do something that unsettles you).

## Flow

Ideally, each of your slides should be logically linked to the one before it, so your presentation flows smoothly. You do not want a scientific presentation that jumps from one subject to another without any logical link. Therefore, when you practice your talk on your own or with peers, it's important to look out for parts that do not flow well. Good flow can be achieved by having a transition statement that links each slide to the next.

## Learning Through Critical Observation

When you attend a scientific conference, it's good practice to think critically about people's speaking styles as they give their presentations. What do you like? What do you dislike? This will help you to identify things you want to incorporate or avoid in your own presentations.

## Contact Information

At the end of your talk, you should include a slide that contains your contact information so that anyone interested in your work can get in touch with you later, such as your name and email address. Be sure to leave the slide up long enough for interested audience members to record your contact information.

## Preparing and Presenting Posters

Another important way of communicating your research at scientific conferences is in the form of a poster (Fig. 9.9). You may feel more comfortable presenting posters at your first couple of conferences, as this provides some presentation practice without the pressure of giving a talk to a large audience. Often, a poster contains similar elements as a scientific manuscript, such as a title, author list, abstract, introduction,

**Fig. 9.9** Presenting posters at scientific conferences is an important way of communicating your research and networking with other scientists

materials and methods, results and discussion, conclusion, acknowledgment, and reference sections. This standard format helps promote effective communication between the presenter and the audience. Like scientific manuscripts and oral presentations, your poster should be designed to let other scientists know several things: What problem did you address? Why is this problem important? What approach did you use? What did you find? What are the implications of your findings? Presenting a poster is a good way of disseminating your research findings and making connections with other scientists.

Typically, your poster will be part of a poster session at a conference where numerous other posters will also be presented. Poster sessions often contain tens to hundreds of posters, depending on the conference size. Scientists attending a poster session look at the different posters and interact with the presenters of those they find interesting. It is therefore important to make your poster visually appealing and engaging so that it stands out from all the other posters alongside it. A typical interaction may last 5–15 minutes, depending on the level of interest. Presenting posters is often a good way of networking with other scientists and increasing your visibility in your field of expertise. It is usually easier to interact closely with other scientists through a poster than through an oral presentation. Ideally, you want to design your poster to attract people to view it and interact with you. You must thoroughly understand all the information on your poster to communicate it concisely and effectively to anyone interested in your work. In the remainder of this section, we provide some tips on how to create an effective scientific poster.

## Poster Dimensions

Before preparing and printing your poster, it is important to find out the required poster dimensions from the conference organizers. The dimensions of a poster at a conference can vary, but a commonly used size is 48 inches (122 cm) in width and 36 inches (91 cm) in height. The poster dimensions determine how much space you have to include all your information and ensure your poster fits in the space provided by the conference organizers.

## Poster Design

Scientific posters are typically designed to have a specific format, such as abstract, introduction, materials, methods, results, discussion, conclusion, acknowledgment, and reference sections (Fig. 9.10). However, some of these sections can be omitted or combined. For instance, it may only be necessary to have an abstract or introduction section (rather than both), and the results and discussion sections can be combined. This standardized format is used because it is highly effective in organizing the information arising from scientific studies and because other scientists are familiar with it and expect it. However, you do not necessarily need to follow this conventional format. It may be desirable to use your creativity to present your findings in a more innovative and compelling fashion, as long as the audience still finds the poster engaging and informative. Indeed, this is one way you can make your poster stand out from all the others at a conference. If you do use the traditional format, then people will mainly focus their attention on the title, abstract, figures, and tables of your poster (rather than the text in the other sections). Consequently, it is especially important to make these sections clear and compelling.

## Title

As well as appearing on your poster, the title of your research may also appear in the printed and online versions of the conference program. You should therefore choose a title that accurately reflects the content of your research but is also attractive to other scientists attending the conference. As with scientific talks, knowing the audience is important when selecting a suitable title. The more general the audience, the more general your title should be. Typically, you want to use a large font for your

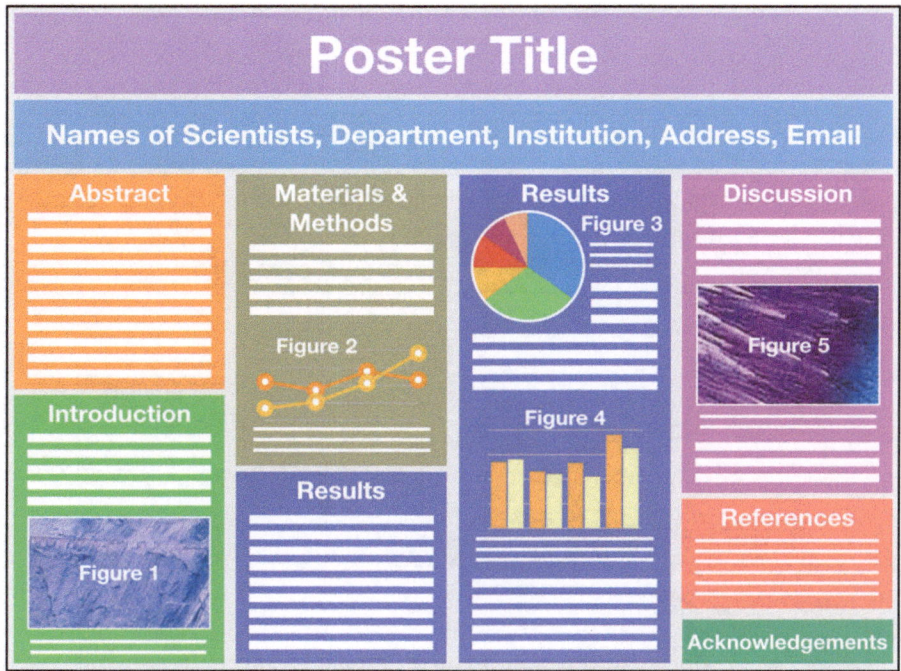

**Fig. 9.10** Scientific posters are often designed to have a specific format, including abstract, introduction, materials and methods, results and discussion, conclusion, acknowledgment, and reference sections. Diagram reproduced from "Scientific Posters: A Learner's Guide" by Ella Weaver; Kylienne A. Shaul; Henry Griffy; and Brian H. Lower (Creative Commons 4.0). However, it is sometimes advisable to use your creativity to create a poster with a more innovative style to distinguish it from other people's work

**Table 9.1** Recommended word counts, image counts, and font sizes for a scientific poster. Typically, you want to use only one or two different font styles in your poster

| Poster section | Word count | Number of figures or tables | Font size |
|---|---|---|---|
| Title | 8–15 | None | 60–120 |
| Authors/Institutions | 25–200 | None | 60–80 |
| Abstract | 50–100 | None | 30–36 |
| Introduction | 50–150 | 1–2 | 30–36 |
| Materials and Methods | 50–150 | 1–3 | 30–36 |
| Results and Discussion | 50–150 | 1–5 | 30–36 |
| Acknowledgments | 10–100 | None | 24–30 |
| References | 25–200 | None | 24–30 |

title and author names so that people attending the poster session can easily locate your poster and learn what it's about (Table 9.1, Fig. 9.11, bottom). You may also want to include the logo of your institution next to the title to clearly show where the research was performed.

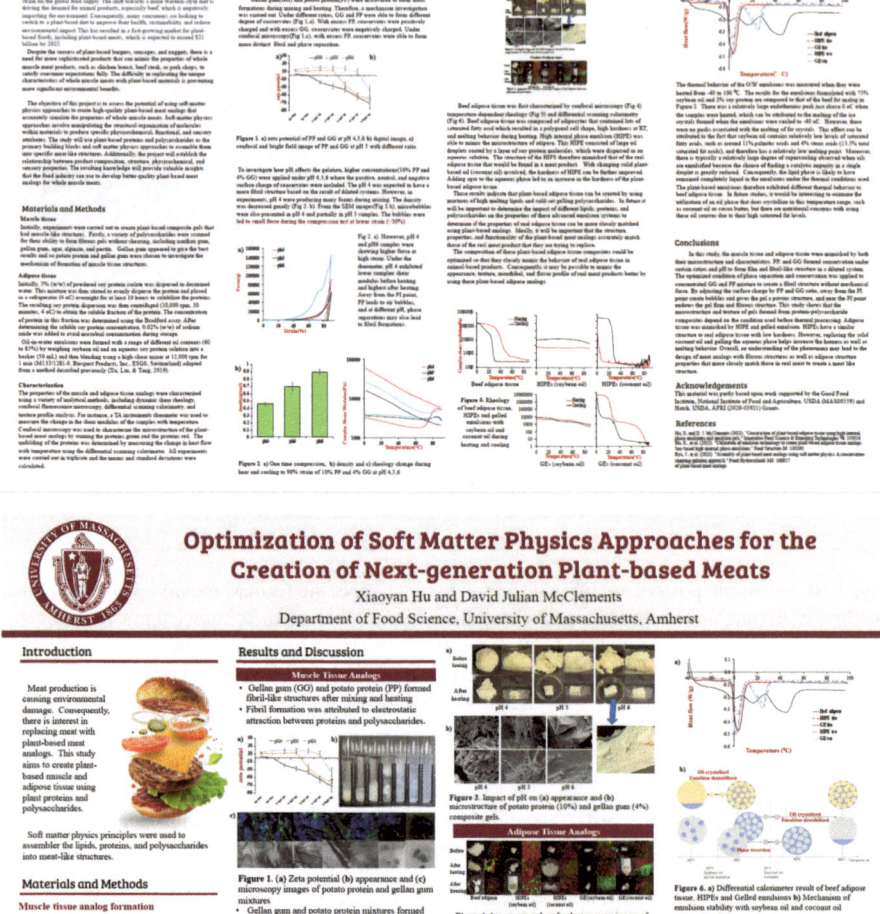

**Fig. 9.11** Examples of a visually unengaging (top) and a visually engaging (bottom) poster. It is better to include more diagrams and figures in your poster rather than too much text, as the audience will find this uninteresting

## Text and Images

When designing a poster, people often want to include as much information as possible in the limited space available. As a result, they put far too much text into the poster, which is often too small to easily read (Fig. 9.11, top). Typically, viewers will be put off by a word-dense poster and will not have the time or inclination to read it. It's much better to include less text and more images, as this will make your poster visually attractive to potential viewers (Fig. 9.11, bottom). For instance, in the *Materials and Methods* section, it is better to include diagrams of your experimental methods, such as images of equipment and protocols, rather than describing all the methods in detail using text. In your *Results and Discussion* section, you could include diagrams of any physicochemical or biological mechanisms you propose to interpret your results, as this will make it easier to explain difficult or important concepts to anyone who asks you questions. The figures and tables should be designed to be large enough to view easily, as well as being visually appealing. The text should be large enough to read, and the combinations of colors used should be impactful. A general rule of thumb is to use a font size that can be read from approximately 3 feet (1 meter) away, as this is the distance a person will typically stand when viewing your poster. You can use the figures, tables, and diagrams in your poster to explain any details that might be important to someone who is interested in your work. Nevertheless, the contents of your poster must accurately reflect the research you have carried out. Consequently, it's crucial to design a poster that is both visually appealing and appropriately summarizes your research findings.

## Contact Information

It is good advice to provide contact information on your poster so that people can find out more about your research or contact you later. This may include your name, photograph, address, email address, website, or social media handles (such as LinkedIn). Alternatively, you could include a QR code that directs the reader to a paper, website, or personal page. You could also increase the potential impact of your work by bringing single page copies of your poster so that people can take it away with them.

## Poster Design

Your poster should be designed to be visually engaging and easy to understand (Fig. 9.11). You want to select a color scheme that attracts readers. Some universities have templates that can be used for posters that include a color scheme consistent with the institution's logo. Often, there is a color strip at the top and bottom of the poster (to make it visually appealing) with a white color in the middle (to save ink when printing). It's not a good idea to have a colored background throughout your poster, as this will use lots of ink when you print it, which is time-consuming and expensive (especially if you are paying for the printer supplies). The color of the headings and main text should be consistent with the color strips. As mentioned earlier, your poster should have a good balance of text and images. You should also organize the different elements of your poster logically so that viewers can easily follow the progression of your research (Fig. 9.10). In summary, the main attributes to consider when designing an effective scientific poster are as follows:

- *Standardized format*: Title, Authors, Abstract, Introduction, Materials and Methods, Results and Discussion, Conclusions, References, and Acknowledgments.
- *Strong visual impact*: Clear organization, good color contrasts, simple and readable text, and effective use of images.
- *Contact information*: Name, address, website, or QR codes can be included so the audience can find out more or contact you later.

## *Level of Explanation*

You should be aware that viewers of your poster may have different levels of interest. Some people may be working in a very similar area and want to know many details about your work. Other people may work in a different area and only have a mild interest in what you have done. Ideally, you should therefore, not start with a highly detailed explanation of your work to every person who views your poster. It is good advice to start with a brief overview of your poster that does not contain too much detail to gauge the level of interest. You can then ask the listener if they want more details about any parts of your poster.

## Concluding Remarks

Presenting your research findings at scientific meetings is an important way for other people to learn about what you have done. Moreover, it is a great way of increasing your visibility in your field of expertise, as well as building skills and confidence in public speaking and presenting that will be highly valuable for many career paths. It is therefore critical to create oral presentations and posters that are impactful and memorable, as this may help you to find collaborators or future employers. In addition, people who learn about your work at conferences may read and cite your scientific publications, thereby further increasing your impact and standing. An important skill of a successful scientist is therefore to be able to design and deliver impactful oral and poster presentations. You should therefore take as many opportunities as you can to polish your presentation skills during your graduate studies.

**Key Points**

- Effective communication is a crucial skill for research scientists. You need to communicate your research findings at lab meetings, graduate seminars, scientific conferences, industry meetings, and for your final thesis presentation.
- The two major types of scientific talks are research and overview talks. Research talks focus on research carried out by the presenter, while overview talks review a particular topic.
- Good preparation is key to giving an impactful talk. Some of the important steps in creating a presentation include topic selection, title selection, synopsis, topic familiarization, talk structure, slide preparation, and practice.
- Your talk should be informative, engaging, and memorable. The ABT method can be used to structure your talk, which introduces the context, highlights a problem, and outlines the resolution.
- It is important to know your audience when preparing a talk, as this will impact the content and design of your presentation.
- The attention of an audience is usually greatest at the beginning and end of a talk. Thus, it is important to design your talk to have the most impact in the introduction and conclusion sections. A presentation can be designed to contain several themes with their own mini-introductions and mini-conclusions to maintain audience attention.
- An oral presentation is structured differently from a written manuscript. It should include introductory remarks, main body, and concluding remarks. Introductory remarks highlight the relevance of the research and the approach taken. The main body presents the research findings. Concluding remarks include a summary, implications, and future directions.
- Good images are critical for creating impactful presentations as they increase audience engagement and understanding.
- Some practical considerations to consider include timing, aesthetics, slide clarity, flow, and presentation style.
- Posters are a popular means of communicating research at scientific conferences. Posters should be designed to be visually engaging and easy to understand. They usually require a standardized format. A strong visual impact can be obtained by using clear organization, good color contrasts, simple and readable text, and an effective use of images.

## Resources

Anderson S (2016) Ted talks: the official TED guide to public speaking. Mariner Books, Boston

Olson R (2018) Don't be such a scientist: talking substance in an age of style, 2nd edn. Island Press, Washington, DC

Weaver E, Shaul KA, Griffy G, Lower BH (2023) Scientific posters: a learner's guide. The Ohio State University Pressbooks, Columbus

# Chapter 10
# Reaching the Finish Line: Writing and Presenting Your Thesis

## The Thesis

This chapter is mainly targeted at PhD students who need to write a thesis to graduate, but it also contains useful information for undergraduate or Masters students
who may also have to write a thesis. Typically, the final part of your graduate studies
involves writing and presenting your thesis. The thesis is a written document that
contains the main findings of your graduate studies. The number of pages in a PhD
thesis varies considerably depending on the requirements of your institution, your

field of study, and the nature of your specific research project, but it is usually between about 150 and 300 pages. It is important to note, however, that the quality of the research reported is usually more important than the number of pages. It is possible to write a short thesis that contains cutting-edge research or a long thesis that contains little value.[1] The main goal of your thesis is to show that you have contributed new knowledge to your field of study and have demonstrated the skills needed to be a productive and successful scientist. A PhD thesis typically has a standardized format that consists of several main sections (but all of these are not always required):

- Title page
- Acknowledgments
- Abstract
- Table of contents
- Introduction
- Background literature
- Main body of research chapters
- Conclusions
- References
- Appendices

A PhD thesis usually follows this standard format because it helps examiners and other readers easily locate relevant information. In some institutions, the format may differ from this one. For instance, some universities allow students to include published manuscripts as part of the main body of their thesis, without having to reformat them. In this situation, it is much easier to write your final thesis if you have already published one or more manuscripts. It also has the added benefit that the manuscripts have been reviewed by experts in the field and the science reported was deemed worthy of publication. It is therefore more difficult for your thesis committee to reject your work.

## When Should You Write Your PhD Thesis?

Whenever possible, it's a good idea to start planning your thesis during the early stages of your graduate studies and to write parts of it as you proceed through your program. This will make it much easier and less stressful when it comes to writing the final version. Once you have a fairly good idea of the topic you are working on, then you can sketch out the different sections of your thesis. These can change as your research develops, but it is useful to have a general idea about what you intend to do. You could use the structure provided in the following section to do this.

---

[1] It has been reported that the PhD thesis of David Rector, who did his graduate studies at MIT, was only 9 pages long ("An Unstable Adams Spectral Sequence", 1966).

Typically, it is recommended to write a review article toward the beginning of your graduate studies. As mentioned earlier in this book, this helps you to get up to speed with the current state of knowledge in your area of expertise, as well as to identify any gaps or controversies that your research could address. If it's published, this article may also enhance your scientific reputation. Another important reason for writing a review article is that you can use it as part of the background literature chapter in your thesis. It is also recommended to focus on writing original research articles throughout your graduate studies, as each article could be used as a separate research chapter in your thesis. Of course, this depends on your field of study and your research progress. It is much easier to publish research articles in some fields than others. Even if you cannot publish scientific articles during your studies, it is still useful to collect and analyze your data and write up the results for each series of experiments you do (even if they fail). You may then be able to use these results as chapters in your thesis. If you wait until the end of your PhD to write up all your work, it may be overwhelming, and you may forget some of the things you did years earlier (another reason to keep detailed notes!). Moreover, if there is any missing data, you can carry out the additional experiments required when you have all the materials, protocols, and instrumentation working, rather than waiting until the end of your studies when it may be much more difficult.

In summary, it is a good idea to plan the structure of your PhD thesis as soon as you have a good idea of the research project that you will be working on, as well as to write up as much of your research as you can throughout your graduate studies.

## Sections of a PhD Thesis

Typically, a PhD thesis is divided into several sections that follow a fairly similar format. However, the exact thesis format required varies between institutions, so it is important to read the guidelines for your particular institution before writing and submitting your thesis. It is also helpful to get a copy of a PhD thesis of a student that has already graduated, especially if they worked in the same laboratory as you, as this can be extremely useful in finding out how to structure your thesis. Some tips on writing the different sections of your PhD thesis are provided here. Additional information can be found in the chapter on writing scientific manuscripts (Chap. 6).

## *Title Pages*

The title pages of a PhD thesis typically include the following information:

- *Title*: The title of your thesis should be given at the top of the title page. It should accurately reflect the content and scope of the research carried out. Since there may be a variety of topics included in your thesis, it is important to choose a relatively general title that encompasses them all.

- *Student name*: The full name of the PhD candidate who conducted the research and wrote the thesis should be included below the title. In some institutions, this may be proceeded by a statement such as "A Dissertation Presented by…."
- *Institutional information*: The name of the university or other institution where the research was conducted should be included on the title page. You may also have to include the logo of the university or institution here.
- *Degree information*: The name of the degree for which the thesis has been submitted should be included, such as Doctor of Philosophy or Doctor of Science. In some places, institutional and degree information must be given in a specific format, such as "Submitted to the Graduate School of the University of Massachusetts Amherst in partial fulfillment of the requirements for the degree of DOCTOR OF PHILOSOPHY"
- *Date*: The date of submission or defense of the PhD thesis should be given.
- *Department information*: The department or program name under which the research was conducted should then be provided.
- *Supervisor, committee member, and department head names*: The names of the supervisor and committee members that evaluated the thesis and provided guidance and feedback to the student are often included. The name of the head of the department or program under which the research was conducted may also be given. These names usually appear on a separate page, after the first title page.
- *Copyright statement*: Some institutions require a copyright statement to be included in the title pages to indicate ownership and permitted use of the thesis.
- *Additional information*: Any additional information specified by the institution or department, such as disclaimers, funding sources, or dedications, may be included. This usually appears as a separate page after the first title page.

In some institutions, a title page should be signed by the supervisor, committee members, and department head. Consequently, you may need to prepare a copy of the title page that has these names on it and a place where they can sign the document. Some institutions may also require you to include a plagiarism statement, such as "I declare that this thesis has been composed solely by myself and that it has not been submitted, either in whole or in part, in any previous application for a degree. Except where otherwise acknowledged, the work presented is entirely my own."

A (fictitious) example of a PhD title page is shown in Fig. 10.1 to highlight what they typically look like.

## Acknowledgments

The acknowledgments section of a PhD thesis usually follows the title pages. It provides you with an opportunity to express thanks to individuals or institutions that contributed to your graduate research and your academic development. The content of the acknowledgments section is usually very flexible and depends on your

**Fig. 10.1** Example of a PhD title page, which includes the title, graduate student name, institution, degree awarded, date, and department name. The precise format of the title page is often specified in the guidelines of the graduate school at your institution

**CREATION OF PLANT-BASED MEAT ANALOGS USING SOFT MATTER PHYSICS APPROACHES**

A Dissertation Presented

by

DAVID JULIAN MCCLEMENTS

Submitted to the Graduate School of the University of Massachusetts Amherst in partial fulfillment of the requirements for the degree of

DOCTOR OF PHILOSOPHY

April 2024

Department of Food Science

personal preferences. Some of the most common people or institutions included in the acknowledgments section are highlighted here:

- *Supervisors*: You should acknowledge your supervisor(s) and other members of your PhD thesis committee, as they have usually provided you with guidance, support, and feedback throughout your graduate studies.
- *Research collaborators*: You may acknowledge people who collaborated with you during your graduate studies, such as those who provided you with data, technical support, or advice that was critical to the success of your work. This often includes your labmates, as well as any external collaborators.
- *Family and friends*: You may acknowledge the support, encouragement, and patience of your family, partner, and close friends throughout your graduate studies.
- *Writing support*: If someone provided you with assistance in proofreading or editing your thesis, you can thank them in the acknowledgments section.
- *Institutional support*: If you received institutional support that was critical to the success of your research, such as analytical, computer, or human testing services, then you may want to thank them.

- *Financial support*: If your research was supported financially by grants, scholarships, or fellowships, then you may want to thank the relevant funding source. This may include government agencies, nonprofit organizations, industrial sponsors, or academic institutions.

When writing the acknowledgments section, it's important to have a good balance between expressing your thanks and maintaining a professional tone.


## *Abstract*

The abstract of a PhD thesis provides a concise summary of the research performed, as well as the main findings and their importance. The format of an abstract typically includes the following elements:

- *Context*: You should begin your abstract by stating the context of the research. This statement should highlight the problem being addressed and why it is important.
- *Objectives*: You should clearly state the main objectives or research questions that guided your research. This will help the examiners and other readers to understand the purpose and scope of your research.
- *Materials and Methods*: You should provide a summary of the main approaches used to collect, analyze, and interpret the data in your study, which could include key materials, protocols, instruments, and analysis methods.
- *Results and Discussion*: You should then present the main findings of your research and their interpretation. In some cases, the methodology, results, and discussion sections may be broken into different parts, with each referring to a different experimental chapter in the thesis.
- *Conclusion*: You should then summarize the main conclusions that you have drawn from your research. This should include a statement about the broader implications of your findings for your field of study and society as a whole.
- *Originality*: Finally, you should state any original contributions your research has made to the field. This might include any new findings or how your research solved an existing controversy or filled a gap in current knowledge.
- *Keywords*: Some institutions require a list of keywords to be included after the abstract to reflect the main topics covered in the research. These keywords help researchers locate your thesis using online literature search engines, which can increase the visibility and impact of your work.

Typically, your abstract should be written in a clear, concise, and engaging manner. It should provide readers with enough information to understand what you did, what you found, and why it's important, without overwhelming them with excessive detail. The maximum length of the abstract varies between institutions but is typically around 200–500 words. An example of a PhD thesis abstract is shown in Fig. 10.2.

**ABSTRACT**

The production of meat using livestock animals is contributing to greenhouse gas emissions, pollution, and biodiversity loss. For this reason, there is interest in creating plant-based meat analogs to replace conventional meat products. The main goal of this research was to determine whether soft matter physics principles could be used to create meat analogs using proteins, polysaccharides, and lipids extracted from plants. To achieve this goal, three main objectives were carried out: (i) Establishment of structure-function relationships between the molecular characteristics of plant-derived food ingredients and the physicochemical properties of meat analogs; (ii) Formulation and sensory analysis of chicken analogs assembled from plant-derived ingredients; (iii) Comparison of the gastrointestinal fate of real meat and meat analogs using an *in vitro* digestion model. Meat analogs were assembled from plant proteins (soy, potato, and pea protein), polysaccharides (pectin, locust bean gum, and gellan gum), and lipids (corn or flaxseed oil). A variety of physicochemical methods were used to compare the properties of real meat and meat analogs, including dynamic shear rheology, differential scanning calorimetry, textural profile analysis, confocal fluorescence microscopy, and an *in vitro* digestion model. We found that meat-like structures and textures could be created using a simple soft matter physics approach that involved phase separation-shearing-and gelling. However, sensory analysis showed that these meat analogs did not have the correct flavor or mouthfeel, which highlights the need for further refinement of their composition and structure. An *in vitro* digestion study showed that the proteins in the meat analogs were less digestible than those in the real meat, which could have nutritional implications. The information obtained in this study may provide useful insights for formulating better quality plant-based meat products, which may improve the health and sustainability of the modern food supply. This research is the first to show that a simple soft matter physics approach can be used to create meat analogs.

*Keywords*; sustainability; plant-based meat; texture; mouthfeel; digestion

*Context*

*Research Problem*

*Objectives*

*Materials & Methods*

*Results & Discussion*

*Conclusions*

*Originality*

*Keywords*

**Fig. 10.2** Example of a PhD thesis abstract whose goal was to create plant-based meat analogs using soft matter physics principles

## *Table of Contents*

Some institutions require PhD candidates to include a table of contents near the beginning of their thesis. The table of contents provides an outline of the main sections and subsections of the thesis, along with their corresponding page numbers. Readers and examiners can then use the table of contents to easily locate any information in the thesis. The formatting of the table of contents may be stipulated by institutional guidelines, or it may be based on personal preferences. In some cases, a "List of Figures" and/or a "List of Tables" is required, which provide the numbering, title, and page location of all the figures and tables in the thesis. Again, these lists are provided to make it easier for examiners and other readers to rapidly locate the data presented within the thesis. An example of a typical table of contents is shown in Fig. 10.3. A table of contents can often be created using word processing software (such as Microsoft Word), which utilizes the different headings in the document for this purpose.

**TABLE OF CONTENTS**

**Fig. 10.3** Example of what the first part of a table of contents might look like in a PhD thesis

## *List of Symbols and Acronyms*

For some types of graduate research, it is important to include a separate section at the beginning of the thesis that includes a list of the symbols and acronyms used throughout your study. This allows readers to quickly look up the meaning of important or unfamiliar terms.

## *Introduction Section*

The introduction section of your PhD thesis should provide an overview of your research topic, establish the context and significance of the study, and outline the objectives and structure of your thesis. Some of the common elements found in the introduction section are listed here:

- *Background and context*: You should begin this section by introducing the "big picture" problem your thesis intends to address, as well as explaining why it's important to your field of study or society as a whole. For instance, if your thesis was about using soft matter physics approaches to create plant-based foods, you may begin with a subsection that highlights the negative impacts of livestock animals on the environment, animal welfare, and human health. Alternatively, if your thesis was on using nanotechnology to create better drugs, you may begin by highlighting the poor performance of conventional drugs.
- *Research approach*: After identifying the big picture problem, you should highlight the approach you intend to take to address it. For instance, you might include a subsection that discusses how soft matter physics can be used to create meat-like structures from plant proteins and polysaccharides, or how nanotechnology can be used to increase the bioavailability of hydrophobic drugs. As part of these subsections, you could include an overview of the current status of knowledge in the field, as well as highlighting any limitations, gaps, or controversies that exist. You can then state the unique approach you intend to take to tackle the problem and how it will advance knowledge in your field of study.
- *Research objectives*: After you have stated your approach, you should be able to formulate one or more hypotheses that can be tested during your research. Typically, it is advisable to breakdown a research project into several objectives (usually 3 or 4) that can be addressed in different chapters.
- *Literature review*: A literature review may be included as part of the introduction section, or it may comprise a separate chapter. If you have written a review article during your graduate studies, then you could include it as a separate chapter and only have a brief literature review in the introduction section. Your literature review should provide a concise but critical overview of all the relevant literature related to your research topic. You should summarize important theories, concepts, and findings from previous studies that provide the framework for your own study. It's useful to identify any limitations, gaps, or controversies in the literature that your research intends to address. You may also want to provide an overview of the principles behind any key experimental protocols or instrumentation used in your research.
- *Thesis structure*: It is common to end the introduction with a subsection that outlines the overall thesis structure. For instance, you may provide a brief overview of the main chapters in your thesis and explain how each one contributes to answering your research questions or achieving your research objectives. Providing this information can help the reader understand the logical flow of your thesis.

In general, the introduction section should be clear, concise, and engaging. It should provide the reader (including the examiners) with a strong rationale for why the study was carried out and why you used the approach you did. It should also demonstrate that you have a comprehensive understanding of your research area.

## Background Literature Section

As mentioned earlier, you may include a literature review as part of the introduction section, or you may have a separate chapter that is a detailed literature review. The background literature section of your PhD thesis provides a comprehensive and critical overview of the current status of knowledge in the research area. It demonstrates your understanding of the concepts, methods, theories, and studies in your field and provides context for your research. The key elements of a background literature chapter typically include the following:

- *Introduction*: You should begin by introducing the purpose and scope of your literature review, which will be somewhat similar to the introduction section discussed earlier.
- *Selection criteria*: In some fields of study, such as nutritional or medical research, it is necessary to describe the criteria used to select the literature included in your review. These criteria could include the relevance to the research topic, publication dates, methodological rigor, or theoretical significance.
- *Key concepts and definitions*: It's often useful to have a subsection that defines the key concepts, theories, and terms that are central to your research topic and used throughout your thesis. This will allow readers to better understand the information presented.
- *Organization*: Depending on the nature of your research and the available literature, you can organize your literature review either chronologically or thematically. In the chronological approach, you highlight the historical development of the ideas related to your research field, whereas in the thematic approach, you group studies together based on common themes or concepts. The approach you decide to use depends on your field of study and research objectives. Ideally, you should select an approach that makes the information presented most clear, concise, and compelling to the reader.
- *Synthesis and analysis*: Your literature review should not simply report what other people have done. In addition to providing an overview of current knowledge in the field of study, it should also provide a critical evaluation of the methodologies used and research findings, highlighting the strengths and weaknesses of different studies.
- *Gaps and limitations*: The main reason for carrying out research is to advance the understanding of a specific field. Consequently, you need to identify any gaps or limitations in the literature that your research aims to address. You can then dis-

cuss areas where further research is needed due to conflicting findings, unanswered questions, knowledge gaps, or emerging trends within your field.
- *Connection to your research*: You should connect your own research objectives to the background literature you have reviewed. For instance, you could discuss how the literature provides a foundation for your study, supports your research hypotheses, or highlights the need for your research.

In summary, your literature review should demonstrate that you have a strong grasp of current knowledge in your field of study. It should also provide examiners and other readers with a clear understanding of why your research is important. Many of the tips we provided in an earlier chapter on writing review articles (Chap. 6) will also be useful for writing your literature review. Whenever possible, it's useful to include diagrams and figures in your background literature chapter because these help the reader to understand difficult or important concepts, as well as making it more interesting to read.

## *Main Body Section*

The main body section of your PhD thesis typically consists of several chapters that present your research findings. The structure and content of these chapters depend on your field of study and the nature of your specific research project. However, these chapters usually have a somewhat similar structure to that of an original research article published in a scientific journal (Chap. 6). Indeed, if you have published or submitted one or more scientific manuscripts during your graduate studies, you may be able to use them as chapters in your thesis. This can save you a lot of time when coming to write your final PhD thesis. Nevertheless, you may have to check with your advisor to ensure this is okay. If you have carried out most of the research in the manuscript and have made a significant contribution to writing it (*e.g.,* you are the first author), then it may be possible to include the whole manuscript in your thesis. However, you should still include a statement highlighting any important contributions made by other individuals. If you only played a minor role in a scientific manuscript, such as carrying out a few experiments (*e.g.,* you are not the main author), then you may not be able to use the whole manuscript as part of your thesis.

There are several common elements usually included in each research chapter:

- *Introduction*: In this subsection, you should provide an overview of the chapter's content and its relationship to the overall research objectives of your thesis. You should clearly state the research question or objective the chapter addresses and explain how it connects to your broader research goals.
- *Theoretical or conceptual framework*: If applicable, you may include a subsection that provides the theoretical framework or conceptual background that informs the research covered in the chapter. This may include a discussion of any theories, models, or frameworks used to design or interpret your research.

- *Materials and methods*: In this subsection, you should describe the materials and methods used to perform the research described in the chapter. You should explain the rationale for selecting these methods and why they are useful for addressing the specific research question. As with a scientific manuscript, you should provide enough details so that other people can replicate your work.
- *Results and discussion*: You should present the findings of your research in a clear, concise, and organized manner. Typically, your data should be presented in the form of tables, figures, or diagrams because this enhances the clarity and understanding of your work. You should then interpret, analyze, and discuss your findings in the context of the research question you are addressing. Ideally, you should highlight any patterns or trends observed in your results, compare your results to the findings of previous studies, and analyze your results using appropriate theoretical concepts and models. Any potential limitations in your research should be discussed and possible approaches to overcoming them should be highlighted.
- *Conclusion*: Each of your research chapters should finish with a conclusion section that summarizes the main findings and their implications. You should highlight any new, unexpected, or notable results and provide potential reasons for them. Whenever possible, you should connect the findings of each research chapter to the overall objectives of your thesis. The conclusion section of one chapter may be used to provide a link to the next chapter. For example, you may highlight what has been found in this chapter, but that further work is still needed, which will be covered in the next chapter. This helps to provide a consistent narrative that runs through the thesis and makes it easier for the reader to understand.
- *References*: In some cases, it may be advantageous to include the references at the end of each thesis chapter (because they are easier to find), but sometimes it's better to include them all as a final chapter at the end of the thesis (because this saves space).

It is often advisable to write the introduction, materials and methods, results and discussion, and conclusions sections of a thesis chapter in the same format as used to write a scientific manuscript (see Chap. 6). This is because it will facilitate the preparation of an original research article based on your work (if it has not already been published). In addition, it will ensure your work is presented in a professional way that examiners are accustomed to reading.

## Conclusions Section

The conclusions section of your PhD thesis is where you summarize your main research findings, discuss their implications, and highlight areas where further research is needed. Some common elements that appear in the conclusions section are listed here:

- *Restate research objectives*: Typically, you should begin your conclusions section by restating the main research objectives or questions of your study. This will remind the readers (and examiners) of the original purpose and scope of your research.
- *Summarize main research findings*: You should then provide an overview of your main research findings, discussing how they relate to your original research objectives or questions, and how they contributed to advancing knowledge in your research field. You should also state whether your findings support or challenge existing theories or perspectives.
- *Implications*: You should highlight the implications of your research findings for advancing your field of study, as well as any potential practical applications of your research to society as a whole. For instance, if your thesis was on the development of plant-based foods, you could say something like "The results of this research could lead to the creation of higher quality plant-based foods, which would facilitate the transition to a healthier and more sustainable global food supply."
- *Highlight limitations*: You should highlight any limitations in your research, discussing how they have affected your findings and how they can be overcome. For example, your sample sizes may have been too small or nonrepresentative, or you may not have had the analytical instrumentation needed to address a specific question.
- *Point out future directions*: You should highlight areas where further research is needed to advance the field. This may include research to address unresolved questions in your own work or new research directions that emerged from your research.

Your conclusion section should be written in a clear, concise, and engaging fashion. It should synthesize your entire thesis and provide a satisfying resolution to the research questions or objectives outlined in the introduction section. This is likely to be the last thing your examiners will read, and so it is important to leave a strong impression and show your understanding of the research field and how your own work contributes to it.

## *References Section*

The reference section of a PhD thesis, which may also be referred to as a bibliography, contains a list of all the sources you cited in your thesis, including scientific articles, books, book chapters, reports, magazine articles, websites, and databases. It allows readers to locate and verify the sources you used to support your research findings. The format of the reference section may be specified by your institution, so it's a good idea to check the guidelines before completing this section. You could also check the format of the references in a thesis written by a previous member of your laboratory. Each citation typically has a specific format depending on the type of source. For instance, author names, publication year, article title, journal name, volume, issue, and page number range are usually required when citing scientific manuscripts, whereas author names, publication year, book title, publisher name, and publication place are needed when citing books. For other sources of information, different formats are needed. It is important that you use a consistent style for all your citations. As with scientific manuscripts, it is strongly advised to use reference manager software (like ENDNOTE, RefWorks, or Mendeley) to organize, insert, and format your references (see Chap. 6). This will save you a lot of time and energy. The sooner you start using one of these programs, the easier it will be to write and edit your thesis.

## *Appendices Section*

In some cases, you may include one or more appendix sections at the end of your PhD thesis. These sections include supplementary material that supports the information presented in the main body of your thesis. Some common types of information included in appendices are listed here:

- *Supporting data*: You may include supplementary data (raw or processed) in the form of tables or figures.
- *Theoretical equations or computer code*: You may include theoretical derivations of mathematical models or computer code that was utilized in your research.
- *Questionnaires or surveys*: You may include questionnaires or surveys used in your research to provide readers with more information about the specific questions you asked and the responses you obtained.
- *Interview transcripts*: You may include transcripts of any interviews you carried out, so readers have access to the original material your research was based on.
- *Ethical approvals*: If your research involved the use of human participants or animals, you could include copies of the ethical approvals or consent forms obtained from the relevant ethics review board. This shows that your research was conducted ethically and in compliance with established guidelines.
- *Glossary of terms*: If your research involves the extensive use of specialized terminology or numerous acronyms, you could provide a glossary of terms as an appendix to ensure readers can easily understand the material presented in your thesis.
- *Permissions and copyright*: If you included copyrighted material in your thesis, such as images, charts, or lengthy quotations, you could provide documentation of the permissions obtained from the copyright holders.
- *Supporting publications*: You may also include copies of any scientific papers you presented at conferences or published in journals.

Each appendix should be clearly labeled and cited in the appropriate part of your thesis. The appendices should be organized in a logical and coherent manner, with clear headings and subheadings to enable easy navigation by readers.

## Setting Up Your Final Exam

The precise nature of the final exam depends on the country and institution where you carry out your graduate studies. In the US, your PhD thesis is usually examined by a thesis committee, which typically consists of several academics (3–5) from inside and outside your department. Your professor typically decides who will be on your thesis committee, although they may ask for your input. When you are reaching the completion of your graduate studies, you should talk to your professor about when you should take your final oral exam. This will partly depend on your expected completion date and the timeline of the funding used to support your research. Typically, you will need to send your written thesis to your thesis committee a week or two before the oral exam. You or your professor have to arrange a date and time when all the committee members are available. Traditionally, all committee members were expected to meet in person, but it is increasingly common for one or more committee members to join the thesis defense by videoconference. Some committee members allow you to send your thesis to them by email, whereas others require you to give them a hard copy. There may also be some paperwork that needs to be completed before you can defend your thesis. In addition, some institutions require that a memo be posted that gives the title, time, and date of your defense to the department and public. Again, it is important to find out the requirements of your specific institution and follow them closely; otherwise, this may delay your graduation.

## Oral Presentation of Thesis

Due to the wide array of different exam formats, we cannot discuss them all in detail here. Instead, we mainly focus on the typical system employed in the US. However, we are aware that institutions in other countries do things quite differently – so it is important to ensure you know the exam protocol at your specific institution.

One of the last parts of your graduate studies is to orally defend your thesis in front of your thesis committee. In some institutions, your oral exam is open to the public, whereas in others, it only involves your thesis committee. Typically, you must prepare an oral presentation that describes what you did, why you did it, what you found, and why it's important. The duration of a final thesis exam depends on the requirements of your academic institution, the field of study, and the members of the thesis committee. Normally, oral presentations are approximately 30–60 minutes long and are followed by questions from the audience, which may take another 15–60 minutes. If the presentation is open to the public, members of the general audience may ask questions first, and then the audience is asked to leave, and the members of the thesis committee ask questions. If the presentation is closed to the public, then only the thesis committee asks questions.

## *Designing and Presenting Your Talk*

The structure of your oral presentation should be somewhat similar to that of your written thesis. However, it is important that it is also clear, concise, and engaging, similar to other types of oral presentations (see Chap. 9). The main elements of a final thesis oral defense include the following:

- *Introduction*: You should begin your presentation by introducing yourself. You should then introduce the specific topic of your thesis and show why it's important. The ABT (AND…BUT…THEREFORE) approach discussed in previous chapters can help you create a compelling and engaging introduction to your talk. First, you highlight the current state of knowledge in the field of interest (AND). Second, you point out some problem that needs to be addressed, which creates tension and interest (BUT). Third, you state how this problem will be resolved (THEREFORE). As a concrete example, we use the development of plant-based meat analogs using soft matter physics approaches as the subject of a thesis. First, you could state that the global population is growing, and people are becoming wealthier, and this is leading to an increase in meat consumption (AND). Second, you could state that the raising of cattle for meat is leading to greenhouse gas emissions, pollution, biodiversity loss, animal welfare concerns, and safety issues (BUT). Third, you could state that this problem could be addressed by developing plant-based meat analogs using soft matter physics principles (THEREFORE). You could then highlight the current status of knowledge in this field, as well as any gaps or controversies. For example, you could say that previous researchers have shown that proteins and polysaccharides can form fibrous structures and textures under certain conditions, but there has been little work on using plant-derived ingredients for this purpose. As a result, your research was designed to find out whether plant proteins and polysaccharides could be used to assemble meat-like structures and textures using soft matter physics approaches.
- *Research objectives*: You should then clearly and concisely state your main research objectives (typically around 3–5). For a thesis on plant-based meat analogs, this could be:

  - *Objective 1*: Application of soft matter physics approaches for creating plant-based meat analogs.
  - *Objective 2*: Optimization of functional performance of plant-based meat analogs: Chilling, freezing, and cooking properties.
  - *Objective 3*: Sensory analysis of model plant-based chicken analogs.

- *Hypotheses:* You should then state the hypothesis or hypotheses of your thesis. As discussed previously, a hypothesis is a prediction of the anticipated results based on your current understanding of the field. Typically, you will have a separate hypothesis for each research objective, which can also be shown when you begin discussing the experiments for each objective. Each hypothesis should be clear, concise, and testable.

- *Materials and methods*: You should describe the research design, data collection methods, and data analysis techniques used in your study. You should also explain the rationale behind selecting these methods and highlight any challenges or limitations you encountered during your research.
- *Results and discussion*: You should present the main findings of your research using well-designed diagrams, tables, and figures. Highlight any trends observed and provide an explanation for them based on known concepts or theories. You should clearly explain the significance of your results and how they contribute to addressing the research objectives of your thesis.
- *Conclusion*: In your concluding remarks, you should summarize the key findings of your research, highlight how they have advanced knowledge, and summarize their practical significance.
- *Acknowledgements*: Finally, you should end by thanking your advisor, committee members, colleagues, and anyone else who supported you throughout your studies.

Although we have listed the key elements of a final thesis defense above, it is advisable to organize the main body of your oral presentation around different objectives. Within each objective, you can include materials, methods, results, discussion, and conclusions for that part of the research (Fig. 10.4). More information

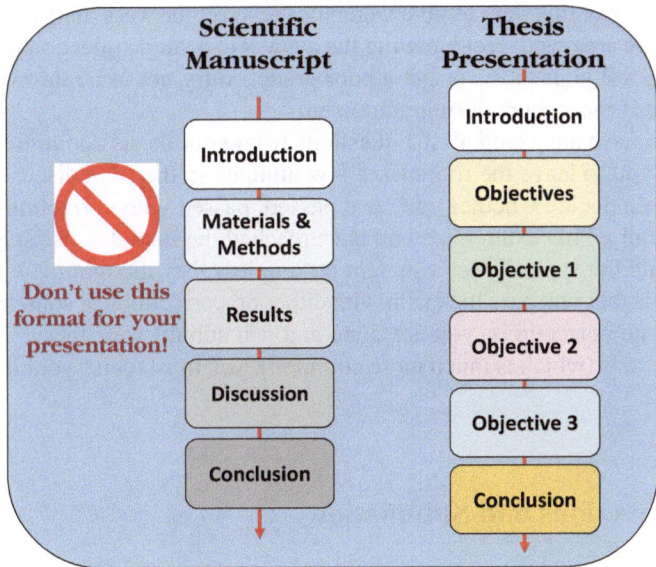

**Fig. 10.4** It's important to design your final thesis presentation to be clear, concise, and engaging. As this is a comprehensive piece of work, breaking the main body of your talk into a series of objectives is usually more interesting to the audience than talking about all the materials and methods, and then results and discussion

about this can be found in the section on theme-based presentations in Chap. 9. This type of organizational structure will help make your presentation more engaging to the audience.

## Answering Questions

Before giving your oral presentation, you should have a deep and broad understanding of the material you are presenting, including the background literature, experimental methods employed, data analysis techniques used, and conclusions drawn. This is important because the thesis committee can ask you any question related to your research, and they expect you to give a reasonable answer. When answering questions, you should give clear and concise responses. Don't waffle or talk about something unrelated to the original question. If you don't understand the question, you can ask the person to repeat or clarify it. If you don't know the answer, you could say so, but try to give some insights into how you might address the question if you had more time or resources. For instance, if someone asked "How do the proteins attach to the polysaccharide molecules in your plant-based meat analogs", you can answer "I am not sure about the exact mechanism of attachment, but it may be possible to find this out by using computer simulations or analytical methods like infrared, nuclear magnetic resonance, or atomic force microscopy." If you have no clue about the answer, it's okay to say so. It's not possible to know everything. It's much better to say this than to give vague or incorrect answers that show you don't understand the area well. Not knowing the answer to a single question is unlikely to cause you to fail your thesis or get a poor grade, so try not to be thrown off or too disheartened if this occurs during your exam

After you have answered all the questions from your thesis committee, they will usually ask you to leave the room for a few minutes so they can discuss your case. They will then decide whether you have passed, passed with corrections, or failed. It is rare to fail a PhD exam once you have reached the final defense stage since an advisor would not typically let a student do it unless they thought they would pass. More likely is that you pass the exam with either no corrections or some corrections. If you have no corrections, you are done and can submit your thesis. If you have some corrections (which is much more common), you must revise your thesis before submitting it.

## Final Corrections and Submission

After you have completed your PhD oral defense, your thesis committee may require you to make some changes to your thesis based on their comments and suggestions. Sometimes, this requires you to carry out more experiments to obtain additional experimental data (although this is rare in many fields). More commonly, you are asked to edit the text of your thesis to correct errors and improve the grammar, redraw graphs or tables to make them clearer, include additional statistical analysis, or reanalyze some of your experiments using a different theory. After you have made the needed corrections, your advisor and/or other thesis committee members may check it. If the changes you have made are accepted, you can submit your thesis to the graduate school. After that, you are done!

## Concluding Remarks

The final stage of your graduate studies typically involves writing a PhD thesis and then defending it in front of a thesis committee. Writing a PhD thesis from scratch can be a daunting and laborious experience. For this reason, it's a good idea to plan your thesis as early as possible and write parts of it as you move through the different stages of your graduate studies. The best way to do this is to write a review article, which can serve as your background literature chapter, and several original research articles, which can serve as chapters in the main body of your thesis. However, your ability to write several scientific manuscripts during your PhD is highly dependent on your field of study. In some fields, it is normal to publish none or only one paper from a thesis, whereas in other fields, it is possible to publish several papers. Some universities require that students write a prospectus thesis midway through their graduate studies. This can help with writing the final thesis because it means that you have already completed a substantial part of it. Your oral presentation should accurately describe what you did, why you did it, what you found, and why it's important. It should also be clear, concise, and engaging to the audience. You should therefore design your oral presentation to flow well and be visually compelling. More information on designing good oral presentations was given in the previous chapter. It is tempting to put everything you have done during your graduate studies into your final thesis presentation, but this can be overwhelming for the audience and committee. For this reason, it is usually better to leave some of the information out of your final presentation and explain the remainder more clearly and compellingly.

It is common once you have completed your final defense and submitted your thesis to feel both excited and deflated. You are excited because you have finally finished and achieved a major accomplishment, but you may be deflated because you have dedicated your life to this work for several years and now it's all over. You have to find something else to do with your life, which can be scary. Ideally, you will already have another job lined up, but if you don't, you may decide to take some time out or look for suitable employment. Whatever your situation, it is important to celebrate and be incredibly proud of yourself for completing a PhD as this is not easy and requires serious dedication and discipline!

**Key Points**
- The final part of your graduate studies involves writing and presenting your thesis.
- A thesis is typically written in a standardized format, such as title page, acknowledgments, abstract, table of contents, introduction, background literature, main body, conclusions, references, and appendices.
- Different institutions have different requirements for the thesis format and so it is important to check before writing and submitting.
- Writing scientific manuscripts during your graduate studies, including a review article and one or more original research papers, will greatly help you prepare your thesis.
- The title and abstract of your PhD thesis should be engaging and informative to attract readers.
- Your abstract provides a concise summary of the research performed and the main findings.
- The introduction section provides background and context, research approach, research objectives, literature review, and thesis structure.
- A background literature review may be included in the introduction, or as a separate chapter. It typically includes an introduction, key concepts and definitions, synthesis and analysis, gaps and limitations, and a connection to your research.
- The main body section consists of several chapters that present your research findings. Each chapter should typically include introduction, theoretical or conceptual framework, materials and methods, results and discussion, conclusion, and references sections.
- The conclusions section should restate your research objectives, summarize your main findings, highlight their implications, indicate limitations, and suggest future research directions.
- The appendices section may contain supplementary materials such as supporting data, theoretical equations, questionnaires or surveys, interview transcripts, or supporting publications.
- It is advisable to design the structure for your PhD thesis as soon as you know your research project. It is helpful to write up your research findings throughout your studies.
- In the United States, your thesis is usually examined by a committee consisting of several academics. You typically orally defend your thesis in front of this committee and provide them with a written copy of your thesis.
- Your oral presentation should be clear, concise, and engaging, and should include introduction, objectives, materials and methods, results and discussion, and conclusion.
- You will likely have to answer questions during your defense, so it is important to have a deep and broad understanding of the material you are presenting.
- After completing your oral defense, the committee gives you suggestions to improve your thesis. You may have to perform more experiments or correct the grammar or formatting.

# Resources

Allen JE (2019) The productive graduate student writer: how to manage your time, process, and energy to write your research proposal, thesis, and dissertation and get published, 1st edn. Stylus Publishing, Stirling

Fisher EM, Thompson RC (2014) Enjoy writing your science thesis or dissertation! A step-by-step guide to planning and writing a thesis or dissertation for undergraduate and graduate science students, 2nd edn. Imperial College Press, London

Turabian KL, Booth WC, Colomb GG, Williams JM, Bizup J, FitzGerald WT (2018) A manual for writers of research papers, theses, and dissertations, ninth edition: Chicago style for students and researchers, 9th edn. The University of Chicago Press, Chicago

# Chapter 11
# Securing Funding: Oiling the Wheels of Research

## Introduction: The Importance of Funding

Scientific research is expensive. Funding is required to pay the salaries of the people working on scientific projects and to cover the costs of reagents, supplies, equipment, travel, and publications. Therefore, scientists need to secure funding to support their research efforts. Professors typically receive "start-up" funds when they begin a new position, ranging from tens of thousands to over a million dollars. This funding is expected to last for the first few years of their employment to help them become established. A professor may use this funding to purchase analytical equipment that is critically important to their research, to pay for graduate students or

© The Author(s), under exclusive license to Springer Nature Switzerland AG 2024
D. J. McClements et al., *How to be a Successful Scientist*,
https://doi.org/10.1007/978-3-031-51402-9_11

postdocs, and to purchase any necessary reagents and supplies. However, a professor is expected to support their research program with external funding for the remainder of their career, which may come from government, industry, nonprofit, or philanthropic organizations. Writing grant proposals to secure funding is therefore an integral part of being a successful professor. It may also be important for graduate students and postdocs, who may have to find funding to support themselves. Moreover, some universities require graduate students to write a grant proposal as part of the examination process. In my department (DJM), we expect students to write an 18-page grant proposal for one of their exams, which they must defend before a panel of professors. Obtaining funding is usually highly competitive. For many federal agencies in the United States, such as the National Institute of Health, the National Science Foundation, or the Department of Agriculture, the funding rates are often less than 10–20%. Consequently, it is critical to write strong grants to be competitive. In this chapter, we give a brief overview of the proposal writing process and provide some tips on how to write successful proposals.

## Identifying Funding Opportunities

Scientists can obtain funding from various sources, including government agencies, industrial companies, nonprofit organizations, and trade organizations (Fig. 11.1). It's therefore important to identify potential sources of funding that may be relevant to your own research. There are various online resources to help researchers identify funding opportunities, which are often made available through academic institutions. Some of the most common online databases and search engines used to find grants are given here:

- *Government grants*: Many governments have online databases where eligible scientists can search for funding opportunities. For instance, the United States government maintains a website known as Grants.gov (www.grants.gov) that

**Fig. 11.1** Examples of different kinds of organizations that provide funding to support scientific research

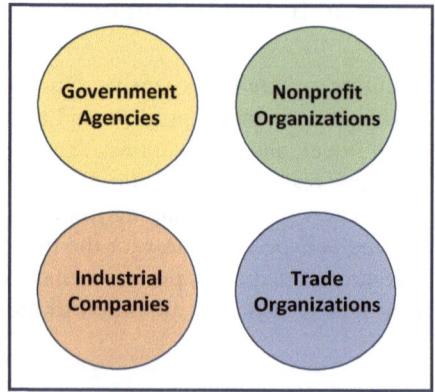

allows researchers to identify funding opportunities from different federal agencies, such as the National Science Foundation (NSF), National Institutes of Health (NIH), Department of Energy (DOE), National Aeronautics and Space Administration (NASA), and United States Department of Agriculture (USDA). These databases can be searched by government agencies, keywords, or categories.

- *Foundation grants*: Numerous private foundations support scientific research, including the Bill & Melinda Gates Foundation, the Howard Hughes Medical Institute Foundation, and the Bezos Earth Fund. Funding opportunities from these organizations can be found using Directory Online, which is a database that provides information on funding opportunities from private foundations and philanthropic organizations (fconline.foundationcenter.org).
- *International grants*: Some organizations have databases to search for funding opportunities related to global issues. For instance, Grantmakers Without Borders maintains a website containing databases of foundations and other organizations interested in supporting projects dealing with internationally important matters (www.internationaldonors.org).
- *Industry grants*: An increasing number of private companies maintain databases where researchers can search for industry funding, such as Halo Science (www. halo.science). These databases provide information about research projects that companies want to fund to address specific problems.

These are just a few examples of databases and search engines that can be used to locate potential funding opportunities. The ones you use will depend on your location and research interests. For most academics, there are one or two federal agencies where they normally obtain most of their funding. For example, one of us based in the United States (DJM) receives most of his funding from the United States Department of Agriculture (USDA) and National Science Foundation (NSF), while another one of us based in the United Kingdom (JM) receives most of his funding from the Engineering and Physical Sciences Research Council (EPSRC). Consequently, knowing the deadlines for submitting grant proposals to these specific agencies is critical. Often, the deadlines for an agency occur around the same time each year. Therefore, you should put this date in your calendar to allow sufficient time to prepare a strong grant proposal well before the deadline. In some cases, a company may contact you directly and ask you to submit a proposal to them because your expertise can address a specific problem they are having. This kind of noncompetitive industry grant is often easier to obtain but has greater restrictions in terms of patenting and publishing and is less likely to occur when you are just beginning to build your academic reputation. It should be noted that many academic institutions allow researchers to subscribe to an email service that highlights funding opportunities available in their field. It is therefore worth finding out if your institution does this.

It is also good practice to communicate with established colleagues in your department or institution who work in a similar area, as they can often give you advice on where to apply for funding to support your research. In addition, most academic institutions have specialized Grants and Contracts Offices that can provide advice and tools to search for funding opportunities.

## Developing a Good Idea for Your Grant Proposal

Before writing a grant proposal, it is essential to develop a well-defined research question with clear goals and objectives (Fig. 11.2). Identifying a good research question depends on several factors. Your research question should address a topic that's a priority to the funding agency. When soliciting proposals, funding agencies typically publish their program guidelines, often called a Request for Proposals (RFP) or a Request for Applications (RFA), which include information about the general areas they intend to fund and the specific topics that are priorities within these areas. These are topics that the agency believes require further research to advance knowledge or address a critical need. You should be sure that your grant proposal fits into one of the priority areas identified by the funding agency. You can have a great idea, but if it doesn't fit, it's unlikely to be funded. Requests for

**Fig. 11.2**  Coming up with novel and relevant ideas is important for writing a good grant proposal

Proposals are typically published on the funding agencies' website and can be found using online search engines. Therefore, you should download this document and carefully read the parts relevant to your grant application before writing your proposal. In addition, your proposed research should be innovative and have the potential to lead to a transformation in science or technology, while also being feasible to achieve with the time and resources available. Consequently, performing a thorough literature search is useful to identify gaps in the current understanding of the proposed topic. Much of the advice given in an earlier chapter for generating ideas when writing scientific papers is also relevant for writing grant proposals (Chap. 6).

## Ensuring You Are Eligible

Before spending time preparing a grant proposal, it is critical to establish whether you are eligible to apply for funding from a particular agency. In some cases, funding agencies have specific requirements about the type of individuals or institutions that are eligible to apply. For instance, the opportunity may only be available for researchers in a particular state, country, or type of institution. Alternatively, it may only be available to early career researchers. The information regarding who is eligible to apply can usually be found within the Request for Proposals.

## Establishing Good Collaborations

In some cases, it is possible to write a grant proposal as a single investigator because you have all the expertise and resources required to accomplish your goals. Typically, these grants are relatively small and focus on a well-defined, narrow research problem. Increasingly, however, grant proposals are being submitted by teams of scientists with complementary knowledge, skills, and approaches (Fig. 11.3). Many current problems in science and society are too large and complex to be addressed by a single person. These problems need people who can bring in expertise and methods from a diverse range of areas. For instance, a large grant proposal on developing plant-based foods could have agronomists, food scientists, research chefs, chemical engineers, psychologists, social scientists, and economists. Sometimes, these teams may have members from one geographical region, but there are also opportunities for obtaining funding for multinational teams, depending on the funding agency. The large and complex nature of the topics addressed by these kinds of grant proposals means that millions or tens of millions of dollars may be requested, which is shared between the different participants. These large grants can help society address important or challenging issues and can lead to significant transformations within a particular scientific discipline.

For multidisciplinary grants, it is critical to establish the right team of people to work on the project. In addition to having all the prerequisite skills, these people

**Fig. 11.3** Forming strong multidisciplinary teams is important for submitting large grant proposals that can address major societal problems

should have the time and commitment to work on the project and the personality to interact with the other team members in a productive and collegial manner. Typically, the principal investigator takes the lead in establishing a team. Initially, they identify who has the required expertise by finding out who has recently published or spoken in the area of interest. The principal investigator may call or email these people to ascertain their interest in the project and establish what contribution they can make. Once a suitable team with the appropriate expertise and experience has been established, there may be a group meeting to discuss the project and clarify individual roles. This could be done by phone or videoconference, but meeting in person is usually more productive. The principal investigator may make an overall plan for the project proposal, assign roles to different team members, and then ask each member to write a particular section based on their expertise. The principal investigator compiles the sections and ensures that the whole proposal is written in a logical and compelling fashion. They may then send it to the other team members for input. After making any suggested revisions, the principal investigator may also send it to one or more scientists not involved in the project to obtain additional insights. The principal investigator therefore plays a role somewhat analogous to being a lead author on a research paper that involves the input of multiple coauthors.

The university where the principal investigator works is typically responsible for submitting the grant proposal. Consequently, they will need to collect all the required documents from each team member, which should be sent well in advance. Often, the team members from other institutions act as subcontractors, which means that the main institution receives funds from the funding agency and then distributes some of them to the other institutions involved. Some larger universities have dedicated staff that can help coordinate the preparation and submission of large grant proposals, which we recommend working closely with if available. Receiving a large federal grant is a major achievement in any academic career and can be extremely rewarding since the investigator has appreciable funding for several years to address a significant problem. However, it is a lot of work to prepare, submit, and manage (if funded).

## Structuring Your Grant Proposal

Funding agencies usually specify the different sections needed in a grant proposal in their call for proposals. The precise sections depend on the funding agency and grant type, so you should carefully read the Request for Proposals document to ensure you include all the sections requested. If you don't include all the needed information with your submission, your proposal may be rejected without being reviewed. Typically, the material that needs to be provided can be divided into two categories: research section and supporting material.

### *Research Section*

The research section describes the rationale, significance, and nature of the intended research. Some common elements in this section are as follows:

- *Summary*: A brief summary of the proposed research is usually needed, often with a prescribed word limit (*e.g.*, <250 words). This summary is used by reviewers to quickly assess the problem the proposal addresses and the approach that will be taken. It may also appear in databases if the proposal is funded so that other scientists, policymakers, and the public can find out what the funded research is about. Consequently, it is crucial to write a summary that accurately and concisely presents the objective, novelty, and relevance of the proposed research. In particular, you want to catch the attention of the reviewers of your proposal as soon as possible. The summary of a grant proposal therefore plays a role somewhat similar to an abstract in a scientific manuscript (Chap. 6).
- *Main body:* The main body of a grant proposal is typically the most important part and is where the researchers state the problem to be addressed, why it is significant, what approach will be used to tackle it, why this approach is appropriate, and why they are the right people to do it. A detailed description of the different parts of the main body section is given later.

- *References*: Any references cited in the grant proposal should be included in a separate section at the end. The most important references in the field of study should be cited, as this demonstrates your knowledge of the area. In addition, you may want to include some of your own relevant references in the grant proposal, as this demonstrates that you have the needed experience and expertise.

## *Supporting Material*

The supporting material provides information that reviewers can use to assess whether the individuals and institutions involved have the experience, expertise, and facilities to successfully perform the proposed research. Some common parts of the supporting material section are as follows:

- *Biography*: Federal funding agencies typically require all key personnel working on a project to provide a brief biography. This includes information about their current position, education, employment, area of expertise, awards, and publication record. Reviewers of the proposal can then assess whether the research team has the experience and expertise needed to successfully perform the research.
- *Conflict of interest*: Federal funding agencies also require applicants to provide a conflict of interest statement for each person working on the project. This typically includes a list of their graduate and postdoc supervisors, students and postdocs they have worked with in their laboratories, and people they have worked closely with in the past few years, such as individuals with whom they have published papers or received grants. This information is used to screen people who may review the grant proposal, since it should not be reviewed by close colleagues or friends of the grant writers, as this may lead to bias and would be unethical.
- *Resources*: Federal agencies usually require a description of the facilities that will be used during the project, including laboratory spaces, library services, computing services, and major pieces of equipment. The reviewers can use this information to ensure that all the resources needed to complete the project are available to the researchers.

## Writing the Main Body of a Grant Proposal

The main body of a research proposal should contain various sections that are usually specified by the funding agency in the Request for Proposals. These include overall objective, hypothesis, specific aims, rationale and significance, research environment, background knowledge, preliminary studies, experimental approach, impact, and timeline. Here, we briefly highlight some of the most important factors to consider for each section.

## *Overall Objective*

Initially, you want to highlight the overall objective of your proposed research and put it into context. You can use the ABT (And... But... Therefore) approach described in earlier chapters to help write a compelling first few paragraphs in your proposal that will grab the reviewers' attention. First, you should provide a brief overview of the current state of knowledge in the field (AND), then highlight where there are significant gaps or controversies that need further research (BUT). Finally, you can state the approach you intend to take to address the identified problem (THEREFORE). To make this approach more concrete, we give an example from a potential grant proposal on developing plant-based meat analogs. "The global population continues to grow, and more people are becoming wealthier, which means that meat production is increasing" (AND). "But, raising livestock for meat is increasing greenhouse gas emissions, pollution, water use, and biodiversity loss, which is damaging the environment" (BUT). "Consequently, there is an urgent need to find more sustainable plant-based food products to replace meat. The overall objective of this research proposal is therefore to use soft matter physics principles to create more sustainable plant-based meat analogs" (THEREFORE). This approach creates tension and then resolves it, which should increase the reviewers' interest. Your main aim in this section is to immediately establish the context and importance of the proposed work, in a way that is easy to understand by people who may be outside your field of expertise. Although grants may be reviewed by experts in your field, they may also be reviewed (or at least initially screened) by people outside your field. Therefore, it is crucial that they are written in a way that is easy to understand, compelling, and avoids getting bogged down in technical details and jargon.

This section may also serve as a bridge between the proposal's overall objectives, hypothesis, and specific aims sections, as it provides the intellectual framework and motivation for performing the research.

## *Hypothesis*

Once a significant problem has been identified and a feasible potential solution has been proposed, it must be framed as a hypothesis. This hypothesis should be a clear, concise, and testable statement about what you expect to happen based on current knowledge. For example, it could be something like, "We hypothesize that soft matter physics principles can be used to create meat-like structures and textures in plant-based foods by controlling the structural organization of plant proteins and polysaccharides." This hypothesis could then be tested by comparing the structural and textural attributes of plant-based meat analogs with those of real meat products. Having a strong hypothesis demonstrates to the reviewers that your grant proposal is based on a sound idea that could be rigorously tested if funding was provided.

## Specific Aims

The specific aims section provides a concise summary of the main research activities that will be performed during the project. Typically, it consists of 2–4 bullet points that distill the essence of the proposed research into a few clear and focused statements. This makes it easier for reviewers to quickly understand the main problems or questions that will be addressed in the research and the approaches that will be employed to solve them. As an example, the following specific aims may be used for our proposal on plant-based meat analogs:

- *"Aim 1: Optimization of soft-matter physics approaches for creating plant-based meat analogs*: The potential of soft matter physics approaches for creating meat analogs from plant-based ingredients will be critically evaluated, including (i) heat-induced fibrillation; (ii) controlled phase separation-gelation; and (iii) lipid droplet formation. We aim to create hierarchical structures in plant-based foods from the nanoscale to the microscale that mimic those formed by muscle proteins, connective tissue, and adipose tissue in whole muscle meats. The microstructure of these plant-based meat analogs will be related to their physicochemical attributes: appearance, texture, and water holding. These experiments will provide an understanding of the link between the structural organization of plant-based meat analogs and their physicochemical properties.
- *Aim 2. Optimization of functional performance of plant-based meat analogs*: Plant-based meat analogs are usually chilled or frozen to increase their shelf life and then cooked before consumption. Cooling or heating meat analogs will alter their appearance, texture, integrity, and moisture holding properties. For this reason, we will systematically examine the impact of temperature changes on the properties of the meat analogs and, if necessary, identify effective strategies to improve their functional performance.
- *Aim 3. Development of model plant-based meat analogs*: As a proof-of-concept, knowledge gained from Aims 1 and 2 will be used to create a plant-based chicken breast analog. Plant-based colors and flavors will be added to the product, and then it will be prepared using standardized cooking methods. The physicochemical and sensory attributes of the cooked product will be compared to those of real chicken meat using a nonvegetarian sensory panel."

Your specific aims should follow a logical sequence. Typically, the studies performed in one specific aim should inform those in a subsequent specific aim. However, it is important that the success of one of the later aims is not entirely dependent on the success of an earlier one. For example, if Aim 1 states that you are going to perform experiments to establish whether you can make plant-based meat analogs that accurately simulate real meat and Aim 3 says you are going to compare these meat analogs to real meat using human sensory tests, then if Aim 1 is unsuccessful, you cannot carry out Aim 3. In this case, you want to write Aim 1 to say that you can already make meat analogs from plant proteins and polysaccharides, but their properties must be optimized. Then, you always have something to test in Aim 3, regardless of the results of Aim 1. In general, clear risk mitigation measures should always be included in your grant proposals, so that the research proposed in the latter parts of your proposal can be performed, or alternative approaches are available if needed.

## *Rationale and Significance*

Funding agencies want to support research that they believe will address some critical societal need or lead to transformative scientific or technological advances. Consequently, you must clearly state the rationale behind your proposed research (why it's needed) and its significance to science and society (what are its potential impacts). You want to highlight to the reviewers and funding agency why your research proposal should be funded rather than all the other ones that have been submitted. You should therefore make a compelling case for the scientific, technological, and/or societal significance of your research. This may include any important advances in knowledge, new insights, discoveries, or innovations, or beneficial impacts on society.

As an example, for our meat analog proposal, we could say something like, "This proposal would lead to an improved fundamental understanding of how plant proteins and polysaccharides interact at the molecular level. This knowledge could then be used to rationally design meat analogs with structures and textures more like real meat, thereby facilitating the transition to a healthier and more sustainable food supply." In addition, you should clearly state how your proposed research aligns with the objectives and priorities outlined in the grant call (RFP) of the funding agency. For instance, for our meat analog grant proposal, this might be:

> The market for plant-based meats continues to grow with an estimated value of over $21 billion by 2025. There is, however, still a need for a new generation of plant-based meats that can more accurately mimic the desirable quality and functional properties of whole muscle meats. The soft matter physics approaches developed in this project could lead to the creation of new processing technologies that could be utilized by the food industry to create high-quality plant-based meat analogs. Greater consumer adoption of these products would have benefits for both the environment and human health, as well as stimulating the US economy.
>
> *Relevance to Program Priorities*: This project addresses two of the priorities highlighted in the Novel Foods and Innovative Manufacturing Technologies (A1364) section of the Food Safety, Nutrition & Health program of USDA: (i) Improve our knowledge and understanding of the chemical, physical, and biological properties of novel foods and food ingredients; (ii) Develop innovative manufacturing technologies that improve food quality.

Thus, the rationale and significance section should convince reviewers that your research proposal addresses an important problem with major scientific or societal benefits, and that it is a priority of the funding agency.

## *Research Environment*

The purpose of this section is to inform the reviewers and funding agency that the researchers involved in the proposed project have the appropriate experience and expertise to succeed and that the host institution has adequate facilities. Detailed information about the research environment is normally provided in the supporting material (discussed earlier), especially the *Biography* and *Resourses* documents. For this reason, it is only necessary to provide a concise but compelling description of why you and your team have the abilities required to successfully perform the proposed research. An example of a research environment statement is given here:

> The principal investigators are highly qualified to successfully perform the proposed project by having the complementary expertise needed. Prof. McClements has considerable experience in the design, fabrication, and characterization of soft matter foods from biopolymers and colloids. Prof. Smith is an expert in taste physiology and sensory science. Both investigators have extensive experience managing successful research projects in the past. The Food Science Department at the University of Massachusetts has all the analytical instrumentation, physical infrastructure, and institutional support required to facilitate the success of the project. It was ranked one of the top graduate programs for Food Science in the USA by the National Research Council (Academy of Sciences) and therefore provides a strong educational environment for graduate students and postdocs.

## *Background Knowledge*

Typically, grant proposals include a section highlighting the current status of knowledge in the proposed field of study. This background knowledge provides the context and rationale for the proposed research. It may point out existing gaps in knowledge or controversies that must be resolved. Furthermore, it might briefly review the principles behind the approach the researchers intend to use in their project. The background knowledge section lets the reviewers know that the researchers fully understand the field, have identified any problems correctly, and have a strong appreciation of the approach they intend to use to tackle them. The background knowledge section is typically between one and three pages.

## *Preliminary Studies*

Grant agencies and reviewers want to have confidence that the proposed research has strong potential for success, or if it is a high-risk/high-reward project, it has the best possible chance of succeeding. This is usually demonstrated by presenting data from preliminary studies that the researchers have previously performed, which highlight the feasibility of the proposed study. For instance, in the case of our meat analog proposal, this may be experimental data showing that we can fabricate and characterize foods with meat-like structures and textures from plant proteins and polysaccharides. If a researcher has worked in an area related to the grant proposal for some time, they may already have extensive preliminary data they can report in this section. However, if this is a relatively new area to them, they may have to perform some preliminary studies so they can include the results in the proposal. Strong preliminary studies are critical to a successful grant proposal.

## *Experimental Approach*

The experimental approach is somewhat similar to the *Materials and Methods* section of a scientific paper. It is where the researchers describe the reagents, equipment, and protocols they intend to utilize for the experiments and data analysis. Typically, it's useful to break the experimental approach section into subsections corresponding to an introduction followed by each specific aim (Fig. 11.4). The introduction section is used to provide a brief overview (usually a paragraph or two)

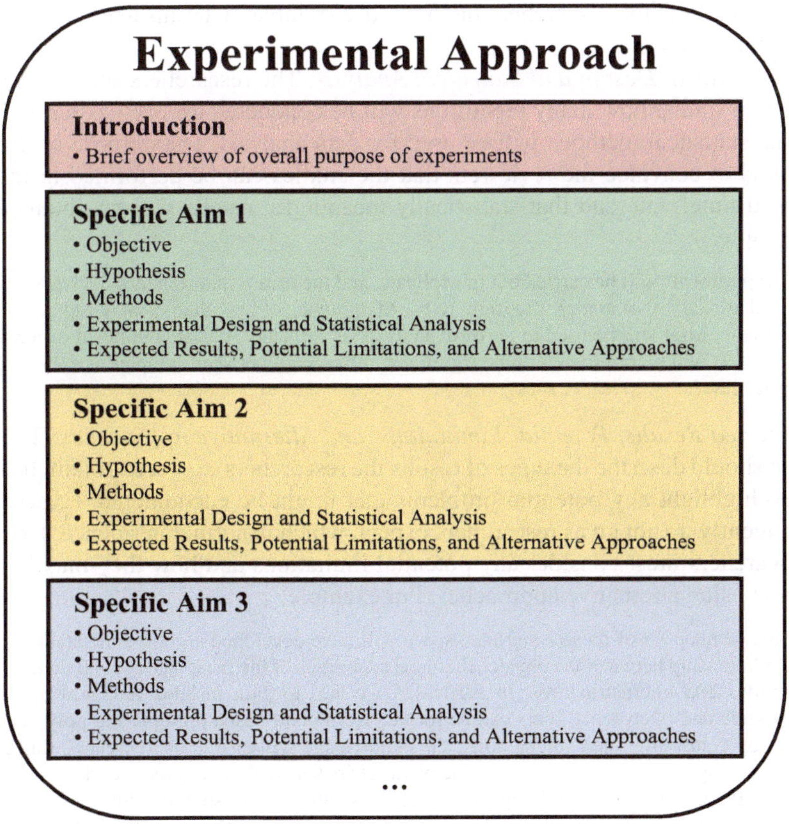

**Experimental Approach**

**Introduction**
• Brief overview of overall purpose of experiments

**Specific Aim 1**
• Objective
• Hypothesis
• Methods
• Experimental Design and Statistical Analysis
• Expected Results, Potential Limitations, and Alternative Approaches

**Specific Aim 2**
• Objective
• Hypothesis
• Methods
• Experimental Design and Statistical Analysis
• Expected Results, Potential Limitations, and Alternative Approaches

**Specific Aim 3**
• Objective
• Hypothesis
• Methods
• Experimental Design and Statistical Analysis
• Expected Results, Potential Limitations, and Alternative Approaches

• • •

**Fig. 11.4** Proposed structure for the *Experimental Approach* section of a grant proposal

of the overall experimental plan that will be used to help orientate the reviewers. Each specific aim subsection is further broken down into a particular format to increase its clarity and consistency:

- *Objective*: Each subsection should begin with a brief description of the overall objective of the experiments conducted for that specific aim. For instance:

  The aim of this series of experiments is to determine whether soft matter physics principles can be used to assemble meat analogs from plant proteins and polysaccharides.

- *Hypothesis:* It is then often advisable to state a testable hypothesis for the experiments in that specific aim. For instance:

  We hypothesize that by controlling the structural organization and interactions of plant proteins and polysaccharides, we will be able to create plant-based materials with meat-like fibrous structures and textures.

- *Methods*: Like in a scientific paper, the reagents, chemicals, equipment, and approaches used in the research should be clearly and concisely presented. It's not possible to provide as much detail as would typically appear in a scientific manuscript. However, sufficient information should be provided to convince the reviewers that the researchers intend to use suitable materials and approaches to reach their aims.

- *Experimental Design and Statistical Analysis*: The researchers should include a section stating how many repetitions will be conducted for each experiment and what statistical methods will be used for data analysis. The purpose of this section is to convince the reviewers that the studies can be performed in the proposed timeframe and that statistically meaningful results will be obtained. For example:

  All experiments will be carried out in triplicate, and the means and standard deviations will be calculated. A statistics program (*e.g.,* Statgraphics Plus, Statistical Graphics Co., Rockville, MD) will be used to perform analysis of variance. A least significant difference (LSD) test will be run to determine significant differences between samples at a 95% confidence level.

- *Expected Results, Potential Limitations, and Alternative Approaches:* This section should describe the types of results the researchers expect to obtain. It should also highlight any potential problems that might be encountered. Experiments frequently do not go as researchers expect, which is normal in science. However, researchers must consider any potential limitations and how they may be overcome using alternative approaches. For example:

  At the *completion* of these *experiments*, we will have developed a good understanding of the relationship between the physicochemical properties of the meat analogs and their compositions and microstructures. In particular, we aim to gain insights into how specific nanoscale and microscale fibers can be formed from plant-based proteins and polysaccharides by controlling their interactions. This knowledge will enable us to create a range of meat analogs with different physicochemical attributes, such as lightness/darkness, hardness/softness, and water holding/expelling. This information should allow us to create chicken, beef, or pork analogs. There are a number of potential limitations to these experiments. First, the meat analogs produced may not have the appropriate appearance, texture, or water holding properties. In this case, we will change the composition or production conditions to alter their properties, *e.g.,* protein concentration, thermal denaturation conditions, and gelling conditions. Second, the initial experiments will only use one type of protein (pea protein) and one type of polysaccharide (citrus pectin). This combination of biopolymers may be unable to produce meat analogs with the desired structures and physicochemical properties. If this is the case, we will examine other combinations of plant proteins (*e.g.,* soybean, lentil, or potato proteins) and polysaccharides (*e.g.,* locust bean or guar gum).

## *Impact*

You should finish your grant proposal with a strong impact statement that clearly demonstrates the importance of the proposed research to the reviewers and funding agency. For instance, you might want to highlight the important scientific or technological advances that will be achieved through the project, as well as the beneficial effects they could have on science and society. It may also be helpful to point out how this research could be used as a platform for future studies. This may be the last part of the proposal the reviewers read; therefore, you want to make a strong and lasting final impression. For instance:

> The proposed research project would generate new knowledge that the food industry could use to fabricate high-quality meat analogs from plant-based ingredients. This research could therefore have a major impact on the environment by encouraging more people to switch from animal-based products to more sustainable and environmentally friendly plant-based alternatives. In the future, we would aim to build on the foundational knowledge gained from this proposal to create a wide range of plant-based foods, such as meat, seafood, egg, and dairy analogs. We would also examine the potential of fortifying these products with micronutrients that might be lacking in a plant-based diet, such as vitamin $B_{12}$, vitamin D, and calcium. We are excited about working on this project and believe it will lead to important new information that will be of great practical relevance to the food industry.

## *Timetable*

It is useful (and sometimes required) to include a timetable for the proposed research project (Table 11.1), which highlights the main activities that will be performed each year. This information is generally presented as a Gantt chart.[1] In addition,

---

[1] A Gantt chart is a widely used standardized graphical depiction of a project schedule. It is a kind of bar chart that shows the expected start and finish times of the different parts of a project. This type of chart is named after Henry Gantt (1861–1919), an American mechanical engineer who originally came up with the idea.

**Table 11.1** An estimated timetable (Gantt chart) to achieve the main objectives of a research project

| Proposed Specific Aims: | Year 1 | Year 2 | Year 3 |
|---|---|---|---|
| **Aim 1:** *Utilization of soft-matter physics approaches to create plant-based meat analogs* | | | |
| **Aim 2:** *Optimization of functional performance of plant-based meat analogs* | | | |
| **Aim 3**: *Development of model plant-based meat products* | | | |

slightly more detail about the proposed activities and their expected duration can be summarized as bullet points in this section. This information helps the reviewers and funding agency assess whether the proposed work is practically feasible within the project duration. This section could also be used to provide a timeline of impact activities, such as publishing work, recruiting students/staff, presenting at conferences, or outreach activities.

## Writing Style

As with other forms of scientific writing, your grant proposal should be clear, concise, and compelling. It should be carefully constructed so that it is easy to read and understand, as well as being visually appealing. The panel members who review your grant proposal may not be experts in your field of study and may have read many proposals in one session. Therefore, you must make your proposal stand out from the others. We recommend including numerous diagrams and figures throughout your proposal to make it more visually appealing and compelling (Fig. 11.5). Furthermore, figures can often help to convey complex experimental protocols or concepts, which would require extensive text to explain and may create issues with word limits.

When writing a grant proposal, it is essential to realize that the first two pages are by far the most important part. If you haven't convinced a reviewer by then that this is an important project that deserves funding, you are unlikely to convince them later. Therefore, it is critical to focus on these first two pages and make them as clear and compelling as possible. For example, presenting the same information with figures and diagrams (Right) is much more engaging than without them (Left).

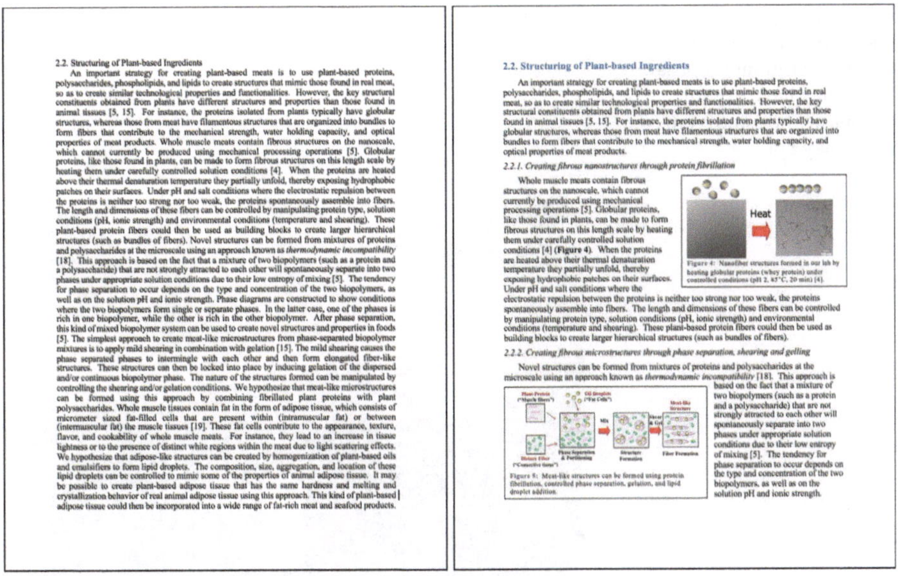

**Fig. 11.5** When writing a grant, it is important to make it clear, concise, and compelling. Including good formatting and many figures and diagrams will help make it more understandable and exciting to reviewers (who may not be experts in your field)

## Submitting Your Proposal

Once you have written your grant proposal and prepared all the supporting documents, you can submit it to the relevant funding agency. If you work at an academic institution, a dedicated Grants and Contracts office usually helps you with the grant submission process. They ensure that all the submission guidelines have been meticulously followed and that all the documents required are included in the final submission. Otherwise, your proposal may be rejected by the funding agency before it's even reviewed, which would be a waste of all the valuable time and energy you have put into preparing it.

## The Grant Proposal Review Process

After you have submitted your grant, the funding agency will usually send it to several reviewers to assess its merit (Fig. 11.6). Typically, a *primary reviewer* reads your proposal in detail, writes a report on it, and presents it at the review panel meeting. This person is generally selected because they have expertise closely related to the subject matter of the proposal (but this is not always the case). There may also

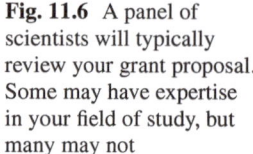

**Fig. 11.6** A panel of scientists will typically review your grant proposal. Some may have expertise in your field of study, but many may not

be several *secondary reviewers* who read your proposal and communicate with the primary reviewer to provide their evaluations. Having numerous people read the same proposal provides a more balanced overall review. Typically, everyone on the review panel meets and assesses the relative merits and limitations of each proposal submitted to a particular program. All the proposals are placed into different categories based on their merits, such as outstanding, high priority, medium priority, low priority, or not funded. The funding agency often specifies standardized criteria that the reviewers should use to evaluate the merits and demerits of each proposal. The specific criteria used depend on the funding agency and grant type. Some of the most common ones are highlighted here:

## Relevance and Impact

- *Relevance*: Does the proposal address a priority identified in the request for proposals (RFP)?
- *Impact*: If the project were successful, would it lead to a significant advance in scientific or technological knowledge or provide an important benefit to society?
- *Dissemination*: Is there a clear plan for disseminating the research findings through publications, presentations, and outreach efforts?

## Scientific Quality

- *Significance*: Is the research question or problem clearly defined, important, and relevant?
- *Innovation*: Does the proposal introduce new concepts, methods, or approaches, or does it build upon existing knowledge in a novel way?
- *Methodology*: Is the proposed experimental design (including equipment, methods, and data analysis) appropriate for addressing the research question?
- *Clarity of objectives*: Are the research objectives, hypotheses, and specific aims clearly stated and aligned with the proposal's overall goals?
- *Feasibility*: Is the research plan feasible within the proposed timeline and budget?

## Research Team and Resources

- *Research team*: Are the principal investigator and other research team members qualified and experienced in the research area?
- *Collaborations*: Are any proposed collaborations likely to enhance the project's potential for success?
- *Resources*: Are the research laboratories, facilities, and equipment appropriate for successfully completing the project?

## Overall Quality of the Proposal

- *Clarity and organization*: Is the grant proposal well organized, with clear writing, appropriate use of figures and tables, and a logical flow of information?
- *Presentation*: Is the proposal visually appealing, with attention to detail and professionalism?

## Ethical Considerations and Compliance

- *Ethical standards*: Does the proposed research adhere to ethical guidelines and standards for scientific integrity?
- *Regulatory compliance*: Is the research compliant with all applicable laws, regulations, and ethical requirements?

The relative weight of these criteria depends on the funding agency and specific grant program, as well as the nature of the grant. A strong grant proposal addresses these criteria effectively, demonstrating the research's significance, feasibility, and potential impact.

This section has mainly focused on the review process of a grant submitted to a federal agency (the most common source of funding in academia). It is useful for researchers to understand this grant review process from the inside. Participating as a grant reviewer can provide valuable insights into the evaluation criteria and expectations of funding agencies, which can help you write better grants. For this reason, it is advisable to volunteer as a grant reviewer toward the start of a career as a faculty member to gain this perspective.

## Acceptance, Rejection, and Resubmission

After the panel has reviewed your grant, it will be given a score or ranking, such as outstanding, high priority, medium priority, low priority, or do not fund. The proposals with the highest rankings will receive funding. The number of successful grants depends on the total amount of funding available. In some cases, your proposal may receive an initial review by a small panel of experts to assess whether it should be reviewed in detail by the full grant review panel. In these cases, the proposals are assessed based on criteria, such as scientific merit, alignment with the funding agency's priorities, feasibility, and overall quality. The proposals that meet the minimum criteria go on for full review, whereas those that do not are rejected. As funding agencies typically receive many grant proposals every call, this is a way of streamlining the process and reducing the workload of reviewers.

After an initial or full review, a funding agency will inform you whether your grant was funded or not. As part of this communication, they will typically include your proposal's ranking and a detailed list of comments and suggestions made by the reviewers. The reviewers will highlight the positive and negative aspects of the proposal.

If your grant is funded, then that's great news! Receiving a positive call from a funding agency is one of the most rewarding aspects of an academic's life. It means you will have sufficient funds and resources to carry out important research for a number of years. Typically, after being informed that your grant proposal has been successful, you will have to complete several pieces of paperwork and update any forms you originally submitted (if any information has changed). The agency then transfers the funds to your institution.

If your grant proposal was unsuccessful, you have to decide whether to revise it and submit it again or simply give up on the idea. Your path forward depends on the comments from the reviewers. When the reviews of your grant proposal are largely positive, and you received a relatively high ranking (such as outstanding, high priority, or medium priority), it's a good idea to revise and resubmit it at the next call for proposals. You should carefully revise your proposal to address all the comments and suggestions made by the reviewers. In some cases, you must include a section in the revised proposal describing how you addressed each reviewer's comments. Resubmitted proposals often have a higher chance of funding because they may already have been deemed high quality by a previous review panel, and you have already addressed any concerns highlighted in your original proposal.

## Tips for Writing Successful Grants

Finally, we summarize tips for writing competitive grant proposals based on those provided by a federal agency within the United States (the USDA).

### *Proposal Development*

- Carefully read the *Request for Proposals* document.
- Note the submission deadline and give yourself ample time to prepare your proposal.
- Develop an idea that fits within the priorities of the program.
- Obtain a successful grant proposal from a colleague as a guide.
- Read the abstracts of recent successful proposals from the same program.
- Obtain critical feedback from your colleagues.

### *Characteristics of Successful Proposals*

- They excite the reviewers.
- They are easy to read and understand.
- They have a clear rationale and objectives that fit the program's priorities.
- They have a clear hypothesis or research question.
- The specific aims are clear and logical.
- They have a strong literature review.
- They include promising preliminary studies that indicate a good chance of success.
- The rationale and significance of the research is clearly articulated.
- The researchers have the expertise required to successfully complete the research.
- Institutional resources are appropriate.
- Clear descriptions of the experimental methods and data analysis procedures are given.
- Expected outcomes, potential limitations, and alternative approaches are presented.
- A good dissemination plan for the results obtained is provided.
- All submission rules have been followed.

## *Characteristics of Unsuccessful Proposals*

- The project has little or no relevance to the program priorities or funding agency mission.
- The scientific quality and/or level of innovation is low.
- The overall objective, hypothesis, and/or specific aims are not clearly articulated.
- Insufficient preliminary data or evidence from the literature is provided.
- The proposal exceeds the page limit or is poorly written.
- The researchers do not have the necessary expertise.
- The institutional resources are inadequate.
- The researchers have a poor productivity record from previous funding (few publications).
- The experimental methods and data analysis procedures are inappropriate.
- The proposal submission rules were not properly followed.

## Concluding Remarks

Scientific research is expensive. Funds are required to pay students and postdocs, to purchase supplies, chemicals, and equipment, to cover travel expenses and to support publication costs. Consequently, applying for funding is a vital skill for research scientists, which requires a combination of knowledge, creativity, writing skills, and planning. However, it is a highly competitive process, with funding agencies typically receiving many more grant proposals than they can fund. Consequently, your grant proposal must stand out from all the others submitted to the same program. This can be achieved by focusing on a problem relevant to the funding agencies' priorities and making sure your grant proposal is presented in a clear, concise, and compelling manner. Persistence and resilience are vital when pursuing research funding, as most grant proposals are rejected. Therefore, it is important to keep refining your ideas and trying to get them funded.

A successful grant application can be highly beneficial to your career. For academics, successful research proposals are listed in your CV, where they provide evidence of your ability to generate innovative ideas and manage a large research program. Moreover, receiving significant funding can increase your publications and citations, as well as the number of students and postdocs you have trained, which can again strengthen your CV. In addition, the research you carry out will advance science and technology in an area that is relevant to society.

**Key Points**
- A critical role for many scientists is to write grant proposals to secure funding for their research.
- Obtaining research funding is extremely competitive and so it's important to write compelling grant proposals.
- You should carefully review any *Request for Proposals* (RFP) documentation before writing your grant and record the deadline for submission in your diary.
- You should identify a research problem the funding agency believes is important.
- The proposed research should lead to important advances in science or technology, or it should address important societal issues.
- The main body of your grant should include several sections (often specified by a funding agency), including objective, hypothesis, specific aims, rationale and significance, research environment, background knowledge, preliminary studies, experiment approach, impact, and timetable.
- You must include all relevant supporting material with a grant application, otherwise, it may be rejected without review.
- After submission and review, your grant proposal may be accepted or rejected based on the reviewers' comments.
- If your proposal is rejected, it's often advisable to revise and resubmit it.
- Securing research funding is important for advancing your academic career.

# Resources

Douglas MR, Bahr M, Olson R (2022) The narrative gym: for science graduate students and post-docs. Prairie Starfish Press, North Haven
Olson R (2015) Houston, we have a narrative. Chicago Press, Chicago
Schimel J (2012) Writing science: how to write papers that get cited and proposals that get funded. Oxford University Press, Oxford

# Chapter 12
# Preparing for the Future: Building a Strong Curriculum Vitae

## Thinking About the Future

Carrying out and completing your graduate studies can be extremely rewarding, but ultimately, it is a means to an end. Eventually, you will finish your academic studies and move on to the next stage of your career. Consequently, throughout your graduate studies, you should consider what you want to do next and how to develop the skills needed to achieve your goals. Moreover, you should focus on

© The Author(s), under exclusive license to Springer Nature Switzerland AG 2024    263
D. J. McClements et al., *How to be a Successful Scientist*,
https://doi.org/10.1007/978-3-031-51402-9_12

engaging in activities to build a strong *curriculum vitae* (CV), as this will help distinguish you from others and improve your employability. In this chapter, we give an overview of some activities that will help you create a strong CV. If you want to become an academic, there are three crucial areas to focus on: research, teaching, and service. Different skills may be more important for other careers, such as business, industry, or government positions. For instance, if you want to work in industry or become an entrepreneur, you may want to take some business or management classes.

## Research Activities

### *Scientific Publications and Presentations*

Publishing manuscripts in academic journals will strengthen your scientific standing, CV, and employment prospects. The higher the number and quality of your manuscripts, the greater the impact. Publishing scientific manuscripts demonstrates that you have the skills to perform important research and prepare it for publication, such as creativity, critical thinking, time management, organization, collaboration, and effective communication. These skills are coveted in academic, government, and industry positions. Being the first author of a manuscript is most desirable, but being a coauthor later in the author list is also valuable. Consequently, you should aim to publish as many first-author and coauthor scientific manuscripts as possible, especially in respected journals (those with high impact factors). It should be stressed that not all PhD studies lead to scientific publications. Some research projects are so challenging that they do not generate sufficient data to write a scientific manuscript. In this case, you cannot include a list of scientific publications in your CV. However, you can include the various skills you learned during your studies, such as analytical instruments, experimental protocols, data analysis methods, and computer software.

Presenting your research findings as talks or posters at scientific meetings also helps to build your scientific reputation and CV. Typically, oral presentations carry more weight than posters, but both are important. You should therefore try to present your work at as many scientific conferences, symposia, and other meetings as possible. However, your professor may only have a limited amount of funding to support your attendance at scientific meetings, so you may only be able to attend a few of them. However, if you volunteer as a leader, organizer, or helper for some professional organizations, they may provide you with financial support to attend their meetings. In addition, there may be opportunities to present your findings at your own institution, such as lab group meetings or seminar series.

## *Awards, Honors, and Scholarships*

Many academic departments and scientific organizations have one or more awards, honors, or scholarships that they periodically (usually annually) give to students. It is therefore advisable to find out about the different ones you can apply for. For example, you could ask your graduate program director or professor about your eligibility for any upcoming awards. For scientific or professional organizations, you could also carry out an online search using appropriate keywords (such as "award," "student," and "American Chemical Society"). You should then carefully read the criteria to determine who is eligible for the award and what documentation needs to be submitted. Typically, this includes a cover letter and CV from the student, as well as one or more letters of recommendation from people who know the student. These letters often come from professors with whom the student has carried out research, in addition to professors with whom they have taken classes. In some cases, organizations require students to complete an application form when applying for a specific award. Receiving awards helps create an impressive CV that distinguishes you from other candidates applying for the same job, scholarship, or (another) award. Furthermore, an award may come with a certificate, plaque, money, or travel expenses to attend a conference (which is nice!).

## *Laboratory Skills*

During your graduate studies, you should aim to develop laboratory skills that are important for your research but may also be useful in your future career (Fig. 12.1). These skills are typically listed in your CV, which helps potential prospective employers determine whether you are suitable for an open position. These laboratory skills include becoming proficient in analytical instrumentation, experimental protocols, data analysis methods, or software programs. For instance, for analytical instrumentation, you could list that you are skilled in gas chromatography, high-performance liquid chromatography, X-ray diffraction, and nuclear magnetic resonance in your CV if you are experienced in using these methods. You could also list the analytical instruments you know with a ranking regarding your level of experience. For example, next to the name of the instrument you could list "expert", "experienced", or "user" to highlight your skill level. It is therefore advisable to develop a range of important laboratory skills during your graduate studies. This often requires you to be curious and proactive. Ask a technician, postdoc, or student to teach you how an instrument works and try to integrate it into your own research work to obtain practical experience and become familiar with its operation.

**Fig. 12.1** It is essential to learn how to use various analytical techniques that will be useful for your future employment

## *Grant Proposals*

If you are hoping to become a professor after completing your graduate studies, it is advisable to gain some experience writing and submitting grant proposals, as this is an essential activity for most academics. Some graduate programs have exams that involve writing a grant proposal and presenting and defending it in front of a team of faculty members. These exams provide graduate students with experience in designing and writing grant proposals. If your graduate program doesn't do this, or you want additional experience, you can ask your professor if they need help writing one of their grant proposals. By working with them, you can get valuable feedback to enhance your grant writing skills. If the grant proposal you helped write is submitted and funded, this will be a valuable addition to your CV because academic departments want to hire individuals who can successfully obtain external funding to support their research programs. Remember grant proposals can be big or small, so while it is important to apply for bigger grants as part of a team, also look out for smaller individual grants that fund activities such as conference travel, hosting summer studentships, and visiting other lab groups. More detailed information about writing grant proposals is provided in the previous chapter.

## *Collaborations*

It is advisable to foster collaborations with other researchers and institutions during your graduate studies whenever possible. Working on collaborative projects helps to broaden your knowledge, expose you to different perspectives, and increase your academic network. Establishing collaborations with people in other institutions may increase your chances of gaining employment with them in the future. People often prefer to hire people they know rather than people they do not. Depending on your field of research, developing collaborative relationships with industry, clinicians, nongovernment organizations (NGOs), and policymakers may also be important.

## *Professional Development*

Universities and professional organizations usually offer in-person or online workshops, seminars, or training programs that allow you to enhance and extend your professional skills. These professional development programs may focus on specific research skills, such as experimental design, data analysis, grant writing, or computer programming. Alternatively, they may focus on other skills important for academic and nonacademic careers, such as teaching, mentoring, time management, equality and diversity, conflict resolution, leadership, and management. Depending on your career goals, it may be a good idea to take some of these courses and list them on your CV. These courses can take time away from your research, so it's a good idea to ask your professor if it's okay before doing them. In some institutions, graduate students are allotted time for self-development activities and actively encouraged to persue them.

## Service Activities

### *Conference Volunteer*

Scientific organizations, such as the American Chemical Society or Royal Society of Chemistry, typically organize one or more conferences annually, where researchers congregate and share their scientific findings. These organizations often need volunteers to help organize and run their meetings. Professors usually take the lead in organizing these events, but students may also be involved. You might therefore be interested in volunteering to help organize or run part of a scientific conference. If you do this, the organization may pay some or all of your travel, hotel, and registration fees. This means you can attend the conference without incurring a large financial cost and include the volunteering on your CV. In addition, you may be able to work closely with some of the professors and other professionals organizing the conference, thereby increasing your visibility and reputation in your field of study.

## Leadership and Organizational Roles

Many academic departments and scientific organizations have student clubs that organize events for their members, such as social events, guest lectures, and tours. These clubs often require a group of people to run them, such as a president, vice president, and treasurer. You might want to volunteer to be part of the leadership team of one of these clubs, as it demonstrates that you have a strong commitment to service/citizenship and have good leadership skills. You may also volunteer to serve in leadership or organizational roles in scientific or trade organizations, as this demonstrates your ability to manage projects and work in a team. Participating in these roles can also increase your professional network, which may be important for your future career (see later).

## Reviewing Manuscripts

Scientific publishers often look for people to review the articles submitted to their journals. You may want to volunteer to be a reviewer for one or more journals in your scientific discipline. Furthermore, your professor may get invited to review articles but not have enough time. Therefore, you could tell them that you would be interested in reviewing manuscripts and when they get an offer to review, they could decline but recommend you instead. Reviewing scientific manuscripts helps you learn what other people are doing in your field and develop a critical mindset for designing, performing, interpreting, and presenting scientific research. It will also enhance your CV. If you review a scientific manuscript, you should keep a record of the journal that you reviewed it for. You can then include the number of papers you have reviewed for different journals in your CV.

## Teaching Activities

If your ultimate goal is to become a professor or teacher, then gaining some teaching and mentoring experience during your graduate studies is a good idea, as this will strengthen your CV.

## *Teaching*

Many graduate programs require students to carry out some teaching activities as part of their degree. Commonly, this involves serving as an assistant in a laboratory, who is responsible for setting up the laboratory (ensuring all the materials and equipment are available), supervising the students during the class, and grading the lab reports and exams. Alternatively, it may involve assisting a professor in teaching a lecture class, such as preparing materials or grading quizzes, papers, and exams. In some cases, it may also be possible to design and teach one or more lectures to undergraduate or graduate students yourself. If you are interested in doing this, you can speak to your professor to see if any teaching opportunities are available. Professors might be traveling during the semester and want someone to teach a class for them while they're away. Alternatively, they may be happy to take a break from teaching for one or more classes or to provide you with an opportunity to gain some teaching experience. You can then list any guest lectures you taught on your CV. Many universities offer courses that graduate students can take to improve their teaching skills. Gaining teaching experience demonstrates your ability to understand and communicate complex concepts to other people, which is beneficial for many academic and nonacademic jobs.

## *Mentoring*

If you are interested in becoming a professor or manager, gaining some experience mentoring other students during your graduate studies is advisable (Fig. 12.2). This could be undergraduate or graduate students working in your laboratory. Mentoring students can be useful for several reasons. It helps you develop communication, collaboration, organizational, and management skills. It also gives the students some research experience, and their results may help you progress your own research.

## **Professional Networking**

Many academic and nonacademic opportunities arise through networking. If someone is looking for a person to fill a particular position or to receive a specific award or scholarship, they are often more likely to give it to someone they know rather than someone they are unfamiliar with. For this reason, it is advisable to establish a strong professional network during your graduate studies. You can do this by connecting with other scientists when attending conferences, workshops, or seminars, especially professors or managers who may be potential employers in the future

**Fig. 12.2** Mentoring other students during your PhD studies can help you build communication and leadership skills

(see Chap. 8). You can also do this by volunteering to serve in leadership or organizational roles at your academic institution or in professional organizations. You should bring well-designed business cards to any meetings you attend where you might meet other professionals in your field.

## Creating a Strong Curriculum Vitae

### *Differences Between a CV and Resume*

The terms "CV" and "resume" are often used interchangeably, but there are some key differences between them. These primarily include their purpose, length, structure, content, and audience.

- *Purpose and length*: A CV is a document that provides a detailed overview of your academic and professional background. It's often used in academic and research settings to highlight your research, teaching, and service activities. It is usually a relatively detailed document consisting of multiple pages. In contrast, a resume is a concise summary of your skills, work experience, and qualifications. It is generally used in nonacademic settings and is tailored to specific job applications. Resumes are usually much shorter than CVs, at only one or two pages long.

- *Structure and content*: A CV typically includes sections on contact information, education, research experience, publications, presentations, teaching experience, grants/scholarships/fellowships, honors/awards, professional affiliations, and referees. Therefore, it provides a comprehensive overview of your academic and research achievements. In contrast, a resume typically includes sections such as contact information, a summary statement, work experience, skills, education, and relevant certifications. It emphasizes the professional experience and skills relevant to the specific job you are applying for.
- *Audience*: As mentioned, CVs are mainly used in academic contexts, such as applying for graduate programs, postdoc positions, faculty positions, research grants, scholarships, or awards. In contrast, resumes are typically used in non-academic contexts, such as business, industry, government, and nonprofit sectors.

Before applying for a job, scholarship, or award, it is crucial to find out whether they require a CV or resume, as well as the precise format needed. Some organizations specify a page length limit and format for CVs and resumes, so adhering to these specifications is essential. We focus on CVs in the remainder of this chapter because they are the most important in academic settings.

## Designing Your CV

It is important for you to have an impactful CV when you are applying for jobs, scholarships, or awards. Many people may be applying for the same position, so your CV must stand out. There are two contrasting aspects to consider. First, your CV should have a standard format because this makes it easier for the people reading it to assess your experiences, skills, and achievements. Second, your CV should be designed to look different from those of other people so it attracts the reader's attention. Your CV should be designed to be accurate, concise, and elegant. You should be sure that there are no typographical or formatting errors by carefully checking it before you submit it. This should be done by you and someone else if they have the time, such as a lab mate or professor, or a family member or friend. Often, someone else can pick up mistakes you miss because you are too familiar with the document. As mentioned, a CV typically has a standardized format that includes several sections:

**Contact Information**

You should start your CV by providing your full name, professional title (if applicable), phone number, email address, and personal website or LinkedIn profile (if you have one).[1]

**Professional Summary and Goals**

You should have a section that provides a concise overview of your background, skills, and career goals. This section should be tailored to highlight your most relevant qualifications for the specific position or opportunity you are targeting. Typically, this statement should only be a short paragraph.

**Education**

You should include a section that lists your academic qualifications in reverse chronological order, starting with your most recent degree. You should include the institution's name, degree earned, field of study, graduation date (or expected graduation date), and final grade (*e.g.*, GPA). Typically, you only include higher education institutions where you got Bachelor's, Master's, or PhD degrees rather than your high school or middle school information.

**Research Experience**

You should include a section that lists your research experiences, such as working in a laboratory for an undergraduate or graduate research project. Each entry should include the title of the project, the main methods used, the main findings, and the dates of the project. If you only have limited space, then you might just include the title and date.

**Technical Skills**

You should include a section that lists any technical skills that align with the position or opportunity you are applying for, such as laboratory techniques, programming languages, data analysis tools, and language proficiency.

---

[1] If you don't have a LinkedIn profile, then we strongly suggest that you create one before applying for positions. You may also consider establishing a profile on other academic networking sites, such as Scopus or ResearchGate.

## Publications

You should include a section that lists any original research articles, review papers, book chapters, or other scholarly works you have published. For scientific manuscripts, you should include the author's names, year, title, journal name, volume, and page range. If the manuscript has been submitted or accepted but not published yet, you could include this information rather than giving the volume and page range, *e.g.*, "(Submitted)" or "(Accepted)."

## Presentations

You should include a section that lists any talks or posters you have given at scientific conferences, symposiums, or workshops. You should include the presenter's names, year, title, event name, location, and date.

## Teaching and Mentoring Experience

You should include a section that lists any teaching or mentoring experiences you have, such as being a teaching assistant, lecturer, instructor, or student mentor. You should specify the courses you were associated with, their level (undergraduate or graduate), and any teaching awards or certificates you received. You should also state the number and type of students you mentored (undergraduate or graduate).

## Grants, Scholarships, and Fellowships

You should include a section that lists any grants, scholarships, or fellowships you received to support your graduate studies. You should specify the funding agency, project title, duration, and amount awarded.

## Honors and Awards

You should include a section that lists any academic honors or awards you received, along with their name, the name of the awarding institution, and the year they were awarded.

## Professional Affiliations

You should include a section that lists any professional organizations, associations, or societies you are a member of. You should also highlight any leadership roles or committee involvement you have had within these organizations.

**References**

Sometimes, employers or organizations require you to provide the names and contact information of several academic or professional references who can vouch for your abilities and qualifications. You should seek permission from your references before including their names in your CV. These references may be contacted by phone or email to provide some background information about you, or they may be expected to write a reference letter and send it to the employer or organization you have applied to.

**Academic Citizenship**

In some countries, candidates are recommended to include information about their academic citizenship activities, such as outreach, committee, and equity, diversity and inclusion work.

Tailoring your CV to the specific requirements of the position or opportunity you are applying for is essential. You should customize the content to emphasize the experiences and skills most relevant to the targeted audience. It is advisable to use CV templates available with word processing programs (such as Microsoft Word). These templates are designed to present your information in a clear, informative, and impactful manner. However, you may want to include some additional elements to make them stand out from other CVs. For instance, you could include a photograph of yourself close to your name and contact information, as well as one or more well-designed images related to the research you have done close to the research experience section. You might also want to ask your lab mates to see their CVs so you can get some ideas about formatting. An example CV is shown in Fig. 12.3 to provide some idea of what they typically look like.

# Concluding Remarks

Graduate studies are an important steppingstone toward your future career. During your graduate program, you should therefore be thinking about what you intend to do in the future and what skills you need to develop to reach your career goals. The most important thing you need to achieve in your graduate studies is to perform good research and write and defend your thesis. However, you can perform various other activities during your studies that can help you develop a stronger reputation and CV. Therefore, whenever possible, you should devote some time to identifying these activities and how you can pursue them without adversely affecting your progress toward your degree. Publishing scientific papers, presenting at conferences, and receiving awards are some of the most important things you can do to improve your reputation and career opportunities. However, various other things are also important depending on the job you aim to do in the future. For instance, if you want to be

# Isobelle McClements

**+1 (123) 123-4567**                    **IzzyMcClements@MadeUp.com**

**Professional Summary:** Strong experience in organic and polymer chemistry. Knowledge of skills required to synthesize and characterize organic polymers and small molecules. Experience with teaching and mentoring students. Goal to become a professor and establish an internationally recognized research laboratory in sustainable polymers.

## Education

**Vassar College | Poughkeepsie, NY     Graduated May 2023**
Bachelor of Arts in Chemistry with Physics Minor
GPA 3.92/4.00

## Research Experience

**Research Experience for Undergraduates at Princeton Center for Complex Materials | June 2022 - August 2022 |** Princeton University, Princeton, NJ
Undergraduate Researcher | Advisor: Dr. Richard Register
**Project:** Determination of Relative Reactivities of Endo-and Exo- 5-hexyl-2-norbornene in Ring Opening Metathesis Polymerization

**Undergraduate Summer Research Institute at Vassar College | June 2021-August 2021|** Vassar College, Poughkeepsie, NY
Undergraduate Researcher | Advisor: Dr. Joseph Tanski
**Project:** Synthesis of the Chiral Cyclic Amines 1-aminoindan and 1,2,3,4-tetrahydronaphthalen-1-amine by Reduction of Imine Precursors with Activated Metal Borohydrides

## Technical Skills

**Certifications:** CITI laboratory safety and ethical conduct training

**Wet Laboratory:** Nuclear magnetic resonance (NMR), Gel permeation chromatography (GPC), Gas chromatography-mass spectrometry (GC-MS), Glove Box, Schlenk Line, Vacuum Oven, Infrared Spectrometry

**Microscopy:** Atomic force microscopy, Scanning tunneling microscopy, Optical microscopy

**Languages:** Spanish (intermediate)

## Publications

McClements, I.F., Wieslera, C.R. & Tanski, J.M. (2023). Crystallographic and spectroscopic characterization of two 1-phenyl-1H-imidazoles. *Crystallographic Communications*, **79**, 678-69.

McClements, D.J. & McClements, I.F. (2023), Designing healthier plant-based foods: Fortification, digestion, and bioavailability, *Food Research International*, **69**, 112853.

McClements, D.J., Newman, E. & McClements, I.F. (2019), Plant-based Milks: A Review of the Science Underpinning their Design, Fabrication, and Performance. *Comprehensive Reviews in Food Science and Food Safety*, **18**, 2047-2067.

## Presentations

**Princeton Center of Complex Materials,** Princeton, NJ, August 2022
Poster: "ROMP reactivity of endo- and exo-6-hexyl-2-norbornene"

**Mid-Hudson ACS Undergraduate Research Symposium,** New Paltz, NY, April 2022
Poster: "Synthesis of the Chiral Cyclic Amines 1-aminoindan and 1,2,3,4-tetrahydronaphthalen-1-amine by Reduction of Imine Precursors with Activated Metal Borohydrides"

## Teaching

**TA General Chemistry Lab** | Vassar College | Poughkeepsie, NY | 2021-2023

**Fig. 12.3** Example of the first page of a CV, adapted from one of the authors' CVs (Isobelle Farrell McClements)

a professor, you may want to develop some experience in grant writing, service, and teaching.

**Key Points**
- During your graduate studies, you should actively work on building a strong CV that is tailored to your career goals.
- If you are interested in pursuing an academic career, your CV should include substantial work in research, teaching, and service areas.
- Research activities in your CV include scientific publications, presentations, awards, honors, scholarships, laboratory skills, grant proposals, collaborations, and professional development activities.
- Service activities include volunteering at conferences, leadership roles, organizational roles, and reviewing manuscripts.
- Teaching activities include mentorship or teaching positions, such as teaching assistantships.
- Professional networking provides opportunities to advance your career. You can connect with other scientists while attending conferences, workshops, or seminars.
- The terms CV and resume are often used interchangeably, but there are some key differences. A CV provides a detailed and comprehensive overview of your academic background, while a resume is a concise summary. A CV is typically used in academic settings, as it gives a detailed overview of your achievements.
- A CV should contain the following sections: contact details, professional summary and goals, education, research experience, technical skills, publications, presentations, teaching/mentorship experience, grants, scholarships, fellowships, honors, awards, professional affiliations, and professional references.

# References

Clay D (2018) How to write the perfect resume: stand out, land interviews, and get the job you want. Independently Published, Amazon

McGrimmon L (2014) The resume writing guide: a step-by-step workbook for writing a winning resume, 2nd edn. CreateSpace Independent Publishing Platform, North Charleston

Singh H, Kothari DP (2019) Written and oral technical communication skills for engineers/scientists. LAP LAMBERT Academic Publishing, Bassin

# Chapter 13
# Final Remarks: How to Be a Successful Research Scientist

## McClements' Top 10 Tips for Being a Highly Successful Research Scientist

In this book, we have offered comprehensive guidance on various topics crucial to being an effective graduate student and research scientist. Within this concluding chapter, we summarize the key elements to provide a reader-friendly resource that encapsulates the most crucial aspects:

D. J. McClements et al., *How to be a Successful Scientist*,
https://doi.org/10.1007/978-3-031-51402-9_13

1. *Understand the system*: Understand the rules and practices at your place of work. What are the expectations? What are the most important things to focus on? What do you need to do to succeed? For example, how many class credits do you need to take, which exams do you need to pass, when should you take the different exams, what format should your thesis be, and how long should your final thesis defense be? Finding a good mentor can often help you navigate the system.

2. *Set clear goals*: Define your research goals and objectives early in your graduate studies, as this will help you remain focused and motivated throughout your work. Regularly refer back to those goals to ensure you are on track and understand where things may need improvement. Be mindful of your progress toward your degree. Give yourself enough time to prepare your final thesis and presentation.

3. *Know your field*: Become an expert in your field of study. Read broadly and deeply. This will help you know the areas that are important to focus on and those that are not. It will also help you design and interpret your experiments, write your thesis, and defend it during your final oral presentation. In general, you should stay curious and continuously learn new things you can apply during or after your graduate studies.

4. *Develop strong research skills*: Focus on developing strong research skills that will be useful throughout your scientific career, including experimental design, data collection, analysis, and interpretation.

5. *Publish and present*: Make publishing scientific manuscripts and presenting at scientific conferences a priority. Publishing manuscripts helps you to disseminate your work, write your thesis, build a strong *curriculum vitae,* and establish your reputation in the field. Ideally, you should write a review article and one or more original research articles. Focusing on writing scientific manuscripts will make your research more efficient and impactful. Presenting your work at scientific conferences as talks or posters will also help disseminate your research findings, develop your communication skills and confidence, enhance your reputation, and extend your network.

6. *Communicate well*: Make writing and speaking well a priority, as this will help you communicate your research findings more effectively and efficiently. You should aim to write and speak clearly, concisely, and engagingly. Make your research outputs stand out from the multitude of others by preparing papers with engaging text, diagrams, and figures, as well as talks with compelling narratives. Make people care about your work.

7. *Develop a strong work ethic and positive attitude*: Cultivate a strong work ethic, as well as a positive and constructive attitude toward your research and colleagues. Perseverance, resilience, and discipline are critical to making meaningful advances in science. These skills are important because scientific research is difficult, and there will inevitably be challenges you need to face and overcome. You will need to remain positive and develop creative approaches to solve the problems you face. It is also important to respect your colleagues by keeping your laboratory a clean and pleasant place to work.

8. *Collaborate and network*: Foster strong collaborations inside and outside your department, as this will broaden your knowledge and experience. In general, take every opportunity to network with other scientists, especially when attending scientific conferences or symposia, as this will help you find future opportunities and employment.

9. *Time management*: Manage your time as efficiently and effectively as possible. Prioritize things that will help you reach your goals (such as learning new skills, submitting manuscripts, and writing your thesis). Avoid time-consuming things that are not useful for your graduate studies (such as carrying out experiments you do not use or spending too much time on committees). Carefully planning your activities will save you precious time and resources.

10. *Prioritize self-care and well-being*: Graduate studies can be demanding, so prioritize self-care to maintain your physical and mental well-being. Take breaks, practice stress management techniques, engage in hobbies, and seek support from peers and mentors. Although being a graduate student can sometimes seem overwhelming and all-consuming, it is ultimately a job. Therefore, remember to make new friends, gain new experiences, and have fun!

We hope this book will help guide you through your graduate studies and to become a more effective and productive scientist. Being a research scientist is one of the most rewarding careers available. It allows you to use your creativity and drive to solve important problems that can improve our world. We hope you enjoy the journey!

# Index